Dionysius Lardner

The Bee and White Ants

Their Manners and Habits

Dionysius Lardner

The Bee and White Ants
Their Manners and Habits

ISBN/EAN: 9783337342401

Printed in Europe, USA, Canada, Australia, Japan

Cover: Foto ©berggeist007 / pixelio.de

More available books at **www.hansebooks.com**

THE

BEE AND WHITE ANTS,

THEIR MANNERS AND HABITS;

WITH ILLUSTRATIONS OF

ANIMAL INSTINCT AND INTELLIGENCE.

By DIONYSIUS LARDNER, D.C.L.,

Formerly Professor of Natural Philosophy and Astronomy in University College, London.

FROM

"THE MUSEUM OF SCIENCE AND ART."

WITH ONE HUNDRED AND THIRTY-FIVE ILLUSTRATIONS.

LONDON:

LOCKWOOD & CO., 7, STATIONERS' HALL COURT,

LUDGATE HILL.

CONTENTS.

CONTENTS.

CONTENTS.

THE WHITE ANTS. THEIR MANNERS AND HABITS.

CONTENTS.

INSTINCT AND INTELLIGENCE.

CONTENTS.

CONTENTS.

Fig. 54.—Uncovered Apiary.

THE BEE.

ITS CHARACTER AND MANNERS.

CHAPTER I.

1. Moral suggested by economy of nature.—2. Antiquity of apiarian researches—Hebrew scriptures—Aristomachus—Philiscus—Aristotle—Virgil. — 3. Modern observers. — 4. Huber. — 5. His servant Burnens—curious history of his blindness.—6. His wife and son.—7. Pursuit of his researches.—8. Structure of insects.—9. Plan of their anatomy.—10. Hymenoptera.—11. Varieties of bees.—12. Hive bee.—13. The queen—her numerous suitors.—14. Her chastity and fidelity.—15. Her fertility.—16. Her first laying.—17. Royal eggs.—18. Royal chamber.—19. Effect of her postponement of her nuptials.—20. The drones.—21. The workers.—22. Structure and members of the bee.—23. Mouth and appendages.—24. Use of proboscis.—25. Structure of tongue.—26. Honey-bag.—27. Stomach.—28. Antennæ.—29. Wings.—30. Legs.—31. Feet.—32. Sting.—33. Organs of fecundation and reproduction.—34. Number of eggs produced by the queen.

1. NATURE offers herself to human contemplation under no aspects so fascinating, as those in which she renders manifest the provident care of the Creator for the well-being of his creatures. The spectacle of infinite wisdom directing infinite power to bound-

less beneficence, never fails to excite in well-constituted minds the most pleasurable and grateful emotions. Such views of Nature are the truest and purest fountains of that reverential love, which so eminently distinguishes the Christian from all other forms of worship.

In the notices from time to time given in this series of the stupendous works of creation presented in the heavens, and of the benevolent care displayed in the supply of the physical wants of the inhabitants, not of the terrestrial globe * alone, but also of the planets,† which, in company with the earth, revolve round the sun, numerous examples of such beneficence are presented. The vast dimensions of these works, as well as the great importance and the countless numbers of the objects to be provided for, leading the mind naturally to expect a system of provisions established on a corresponding scale, their display, while it excites equal admiration and reverence, produces a less intense sentiment of wonder. When, however, we turn our view from the vast works of creation exhibited in the celestial regions, to the more minute ones presented in the organised world around us, our wonder is as much excited as our admiration, at beholding the same traces of Divine care in the economy of an insect, as were observed in the structure and motions of a planet. There are the same infinite wisdom and foresight, the same unapproachable skill, the same boundless goodness directed to the maintenance of the species and the well-being of the individual, as we have seen displayed in the provisions for a globe a thousand times larger than the earth, or for a cluster of worlds millions of times more numerous than the entire solar system, sun, earth, planets, moons, and all! We have thus before us a demonstration that as the most stupendous works of the universe—the expression of whose dimensions surpasses the powers of arithmetic—are not above Divine control and superintendence, so neither are the most insignificant of creatures—whose existence and structure can be made evident only by the microscope—below the same benevolent care.

2. Among the numerous examples, suggestive of reflections such as these, presented by the insect-world, there is none more remarkable than the little creature, to the character and economy of which we shall devote this notice. How true this is, is proved by the examples of those who, in all ages of the world, have devoted their labours to the observation and investigation of its character and habits. In the Hebrew Scriptures numerous allusions to the bee show that, in those remote times, it had already

---

* See Tracts on the Earth, Geography, Terrestrial Heat, Air, Water, &c.
† See the Planets, are they inhabited? the Sun, the Moon, the Stellar Universe, &c.

2

been a subject of attention with the wisest and the best. Pliny relates that Aristomachus of Soli in Cilicia devoted fifty-eight years of his life to the study of the bee; and that Philiscus, the Thracian, passed so large a part of his time in the woods observing its habits, that he acquired the title of AGRIUS. Among his numerous researches in natural history, Aristotle assigned a considerable share to the bee; and Virgil devoted to it the fourth book of his Georgics:—

> " Protenus aërii mellis cœlestia dona
> Exsequar. Hanc etiam, Mæcenas, adspice partem.
> Admiranda tibi levium spectacula rerum,
> Magnanimosque duces, totiusque ordine gentis
> Mores, et studia, et populos, et prælia dicam.
> In tenui labor ; at tenuis non gloria, si quem
> Numina læva sinunt, auditque vocatus Apollo."
>
> GEORG. IV. 1—7.

> " The gifts of Heaven my following song pursues,
> Aërial honey, and ambrosial dews.
> Mæcenas, read this other part that sings
> Embattled squadrons and advent'rous kings—
> Their arms, their arts, their manners, I disclose,
> And how they war and whence the people rose.
> Slight is the subject, but the praise not small
> If Heaven assist, and Phœbus hear my call."
>
> DRYDEN.

3. In modern times the bee has been the subject of the observations and researches of some of the most eminent naturalists, among whom may be mentioned Swammerdam (1670), Maraldi (1712), Ray, Reaumur (1740), Linnæus, Bennet, Schirach, John Hunter, Huber—father and son,—and more recently Kirby, whose monograph upon the English bees may be regarded as a classic in natural history.

4. Among these, the elder Huber stands pre-eminent, not only for the extent and importance of his contributions to the history of the insect, but for the remarkable circumstances and difficulties under which his researches were prosecuted. Visited with the privation of sight at the early age of seventeen, his observations were made with the eyes and his experiments performed with the hands of others ; and, notwithstanding this discouragement and obstacles which might well have been regarded as insurmountable, he continued his labours for forty years, during which he made those discoveries which have conferred upon him such celebrity.

5. Happily for science, Huber, after losing his sight and at the commencement of his researches, had in his service a domestic, named François Burnens, a native of the Pays de Vaud, in Switzerland. Reading and writing constituted the extent of the

education of this person; but nature had bestowed upon him faculties which, with better opportunities, would have rendered him an eminent naturalist. Huber commenced by employing him as a reader.

He read to his master various works on physics, and, among others, those of Reaumur, in which the admirable observations of that naturalist on the bee are so clearly and beautifully stated. Huber soon perceived by the observations and reflections of his reader, and by the consequences he deduced from what he read, that he had at his disposition no ordinary person, and resolved to profit by him. He accordingly procured the means of prosecuting a series of observations on the economy of the bee, with the aid of the eyes, the hands, and the intelligence of Burnens. All the observations of Reaumur were first repeated, and the accordance of the phenomena, as described by Burnens, with those which had been recorded by Reaumur, gave Huber full confidence; and the master and servant, quitting the beaten path, entered upon new ground, and during a period of fifteen years, prosecuted those researches in the natural history and economy of the bee, which, being committed to writing by the hand of Burnens at the dictation of Huber, were published in one volume about 1792, in form of letters addressed by Huber to Bonnet.

6. Soon after this, Huber lost his invaluable colleague, for servant he had long ceased to be. Burnens was recalled by family ties to his native place, where the personal estimation in which he was held caused him to be raised to a high position in the local magistracy.

Previously to this, Huber had the good fortune to consolidate his domestic happiness by marriage. "My separation from my faithful and zealous Burnens," said Huber, "which was not the least cruel of the misfortunes with which I was visited, was, however, softened by the satisfaction which I felt in observing Nature through the eyes of the being who was dearest to me, and with whom I could commune with pleasure on the most elevated topics. But what more than all the rest contributed to attach me to natural history, was the taste manifested by my son for that subject. I explained to him the results of my observations and researches. He expressed the regret he felt that labours which would, as it seemed to him, so deeply interest naturalists should remain buried in my portfolio. Perceiving, meanwhile, the secret repugnance that I felt against the task of reducing them to order, he proposed to take charge of that labour."

7. From that time our great naturalist was again consoled, by having at his disposal two pair of eyes in place of one. The wife and the son, animated by a common enthusiasm, and urged by

4

conjugal and filial devotion, more than compensated for the loss of Burnens; and the observations and researches were pursued with unabated zeal, and were finally collected and published in the second volume, which appeared about 1814, more than twenty years after the publication of the first.*

8. Since any explanation, however popular and familiar, of the economy and habits of the bee, must necessarily involve very frequent references to its structure and organs, it will be convenient in the first instance briefly to explain the terms, by which naturalists have designated its several parts.

The body of insects in general consists of a series of annular segments, so articulated one to another as to allow more or less flexibility. It consists of three chief parts, the *head*, the *thorax*, and the *abdomen*.

The head consists of a simple segment, the thorax of three, and the abdomen of a greater number, sometimes as many as nine. Each segment is distinguished by its ventral or inferior, and dorsal or superior part.

Insects have three pairs of legs, which are inserted in the sides of the ventral parts of the three thoracic segments of the body; and generally two pairs of wings, which are inserted in the sides of the dorsal parts of the second and third thoracic segments, counting from the anterior to the posterior part of the body.

A pair of members, called *antennæ*, are inserted in the sides of the head, varying much in structure in different classes, and in many, including the bee, have the form of slender and flexible horns, consisting of many minute pieces articulated one to another. These are generally presumed to be tactile organs, and are consequently sometimes called *feelers*.

9. This description will be more easily comprehended by reference to the annexed diagram, fig. 1, which may be taken as a general theoretical representation of the structure of an insect.

As here indicated, the three thoracic segments are distinguished as the pro-, meso-, and metathorax.

10. Insects have been classified by naturalists according to the structure of their wings, and the order to which the bee has been assigned, and of which it is regarded as the type, is the *Hymenoptera*, a compound of two Greek words signifying membranaceous wings.

The section or subsection of the order of Hymenoptera, which in its economy and peculiar construction differs most from all other orders of insects, has been designated by Latreille *Mellifera*,

---

* "Nouvelles Observations sur les Abeilles." Paris, 1814.

a Latin word signifying HONEY-GATHERERS; or *Anthophila*, a Greek word signifying FLOWER-LOVERS.

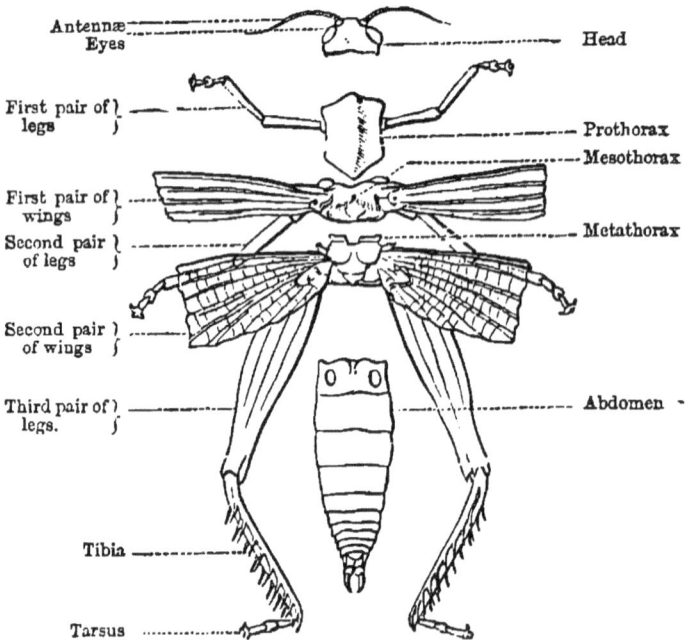

Antennæ · Head
Eyes

First pair of ⎱ · Prothorax
legs ⎰ · Mesothorax

First pair of ⎱ · Metathorax
wings ⎰
Second pair ⎱
of legs ⎰

Second pair ⎱
of wings ⎰

Third pair of ⎱ · Abdomen
legs. ⎰

Tibia

Tarsus

Fig. 1.

11. How numerous are the varieties of bees may be conceived, when it is stated that of bees found in Great Britain alone, Kirby in his Monograph has enumerated 220 species, and other more recent observers have increased the number to 250. The species, however, which by its commercial importance, as well as by its remarkable habits and social organisation, presents the greatest interest, is the Hive Bee, to which, therefore, we shall chiefly limit our notice.

12. The Hive bee belongs to what naturalists have denominated the perfect societies of insects. Each community of these insects consists of three orders of individuals distinguished by their number, their organisation, and the respective share they take in the common labour of the society. These are denominated severally the *queen* or sovereign, the males or *drones*, and the *workers;* the latter consisting of two classes, called the *wax-makers* and the *nurses.* A hive which contains as many as 50000 bees will have only one queen, and not above 2000 males.

13. The *queen* who, as her title implies, is the acknowledged

6

monarch of the hive, is distinguished from her subjects by conspicuous personal peculiarities. Her body, fig. 2, is considerably

Fig. 2.

Queen.

Fig. 3.

Drone.

Fig. 4.

Wax-maker.

Fig. 5.

Nurse, loaded with pollen.

Fig. 6.

Drone in flight, showing organs of fecundation.

longer than that of any of her subjects; she is distinguished by a more measured and majestic gait, by the comparative shortness of her wings, and the curvature of her sting. Her wings, which are strong and sinewy, are only half the length of her body, extending very little beyond the posterior limit of her thorax, while those of the drones, fig. 3, and the workers, fig. 4, cover the abdomen. Her legs are destitute of the brushes and baskets with which those of the workers are furnished. She has no occasion for these instruments of industry, since her exalted station exempts her from labour, all her wants being munificently provided for by her subjects. She is distinguished by her colour as much as by her form, the black of the dorsal part of her body being much brighter than that of the drones and workers, and the ventral parts and legs being of dark orange or copper-colour, the hue of the hinder being deeper than that of the other legs.

The queen, who is the only lady of the hive, enjoys the privilege of being followed by many hundred suitors in the persons of the drones. At the early age of two or three days she is marriageable, and it rarely happens that her royal decision is long postponed; and, indeed, if she were not favourably disposed for such an event, the anxiety of her numerous subjects would urge

7

her to it, for in no human monarchy are the hopes of succession so anxiously cherished as in the Empire of the Hive.

14. It must not be imagined, that because a lady is thus domesticated alone with so many hundred lovers, there is any the least degree of laxity in the morals of the society; on the contrary, although she is absolutely uncontrolled, and is courted by so many hundreds, her choice is strictly limited to one. A fine warm sunny day is selected for the nuptials, which are celebrated in the air. On the auspicious occasion, her majesty issuing from the hive followed by the multitude of her suitors, rises in the air, where she is encircled by the flight of the candidates for her favour. Here she makes her selection, but, alas! the felicity is brief, for the object of her choice never outlives the wedding-day. She is, however, not the less faithful to him, and never contracts a second marriage.

15. Though her majesty is thus left a widowed bride, in two days after the celebration of her nuptials and the loss of her lord, she commences to lay eggs from which a posthumous progeny of that lord, countless in number, are destined to issue. Of the hundreds of rejected suitors, a limited number emigrate with the successive swarms, which from time to time leave the overpeopled hive. Those which remain, being no longer useful to the community, become objects of general aversion, and are finally exterminated by a general massacre, as will presently be more fully explained.

16. During six or eight weeks the queen constantly lays eggs, from which working bees only are destined to issue. Chambers have been previously prepared for these, suitable to the future young ones, in form, size, and position, by the workers. In each of those cells the queen deposits a single egg.

At a later period her majesty begins to lay another kind of egg, from which males will issue. For these also special chambers have been provided by the careful workers, of suitable dimensions, being somewhat more roomy than those prepared for worker-eggs. The number of these male eggs and of the cells for their reception is incomparably less than those of the workers; less, in short, in the proportion in which the drone class is less numerous than that of the workers in the population of the hive.

17. In fine, the queen, sensible of her mortality, and moreover of the approaching state of superabundant population in the hive, lays a certain small number of royal eggs, from which as many princesses issue, who are severally destined to be candidates for the thrones of the colonies which are to emigrate, or to succeed to the throne of the hive itself, should the queen-mother, as often

happens, decide on abdicating and accepting the allegiance of one or other of the emigrating colonies.

18. Special chambers of exceptional form, position and magnitude have been previously prepared for these royal eggs by the provident workers. In these the princesses are reared and educated with extraordinary care, being fed with a peculiar food.

19. It is essential to the prosperity of the community, that the nuptials of the queen should not be postponed to a later period than the second day of her age, the consequence of such postponement being that her progeny would consist of a redundancy of drones. Thus, if the marriage be postponed till she is about a fortnight old, she will lay as many drone as worker-eggs, and if it be delayed until her age is three weeks, she will only lay drone eggs. How great a calamity such events must be in the apiarian economy will be understood, when it is considered that in a well-regulated society there ought to be about ten workers to each drone. The general duration of the life of a queen is from five to six years.

20. The males or drones, fig. 3, are less than the queen and larger than the workers, fig. 4. The extremity of the body is more velvety. The last segment being fringed with hair, extending over the tail, so as to be visible to the naked eye. They take no part whatever in the labours of the community, contribute nothing to the common stock, are idle, slothful, and cowardly, and, as if to render their extermination more easy to the industrious part of the population, nature has given them no sting. They make a louder buzz with their wings in flight, never exercise any industry, and are destitute of the baskets and other appendages with which the busy workers collect the materials of honey and wax.

The life of a drone does not exceed a few months, and he seldom dies a natural death. If he is honoured by the choice of the queen and elevated to the rank of king-consort, he dies on the very day of the nuptials. If he be among the hundreds rejected by her majesty, and do not emigrate with one or other of the swarms, being a useless and idle member of the community, he is massacred by the workers.

21. The workers, sometimes called neuters, are generally considered as sterile females. The number of these in each community is very variable, being seldom less than 12000, more generally amounting to 20000, and in hives where swarming is checked by affording abundance of room, the number may rise to 60000. They are the smallest members of the society, fig. 4, have a long flexible proboscis and legs of peculiar structure.

22. Among the wonders presented by the insect-world the *head* of the bee and its appendages command especial attention.

9

In common with insects generally, the chief parts of the mouth are, the *tongue*, the *jaws*, the *lips*, and the *throat* or œsophagus. The jaws are each double, separated by a vertical division. Each pair opens, therefore, with a horizontal instead of a vertical movement like the human jaws. The pair of upper jaws are called *mandibles*, and the lower *maxillæ.* The upper lip is called the *labrum* and the lower the *labium.* The mouth is also supplied with two pairs of special organs called *palpi* or feelers, one pair attached to the lower lip and called *labipalpi,* and the other to the lower jaw and called *maxipalpi.*

23. In fig. 7, is given a magnified view of the buccal apparatus of the wild bee (*Anthophora retusa*),* the parts being indicated.

Fig. 7.

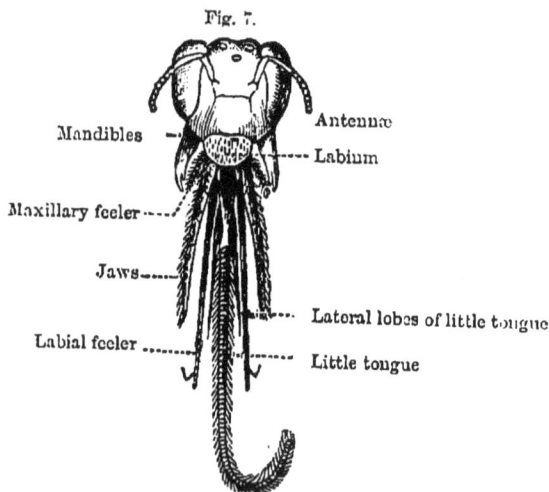

A less detailed view, also magnified, of the same apparatus of the hive-bee is shown in fig. 8.

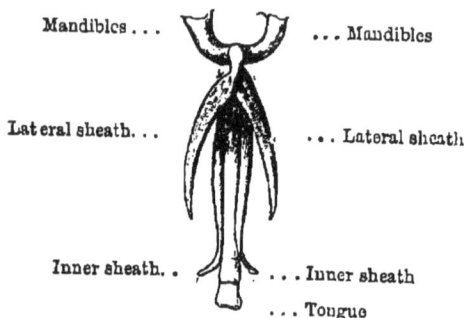

Fig. 8.—Tongue of Hive bee (magnified).

* Milne Edwards.

A magnified view of the head of the drone is shown in fig. 9.

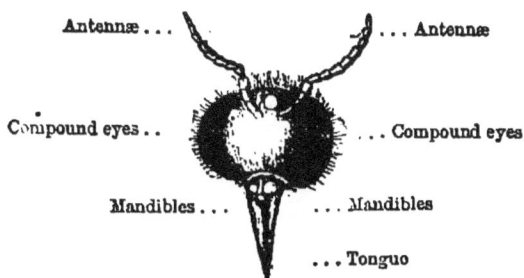

Antennæ . . .   . . . Antennæ

Compound eyes . .   . . . Compound eyes

Mandibles . . .   . . . Mandibles

. . . Tonguo

Fig. 9.—Head of a Drone (magnified).

The mandibles, or upper pair of jaws, in the workers are strong, horny and sharp. They are the tools with which it performs its various labours. Meeting over the other parts of the mouth, they are covered in front by the labrum or upper lip. The maxillæ, or lower jaws, on the contrary are pliable and leathery, and hold the objects upon which the insect works with its mandibles.

The tongue, which is long and endowed with great flexibility, is moved by a complex system of powerful muscles. When it is in a state of inaction, it is withdrawn within its sheaths, the end which protrudes beyond them being doubled up under the head and neck, the sheaths consisting of two pair of strong scales.

24. When the bee lights upon the blossom of a flower from which it desires to extract the nectar, it darts out its tongue from the sheaths that invest it, and having pierced the petals and stamina where the treasure is hidden, it inserts its tongue which moves about in every direction in virtue of its great flexibility and muscular power, and probes to the very bottom the floral cells, sweeping their surfaces and draining them to the last drop of their precious juice. Having thus collected the nectar upon the tongue, that organ being drawn back into the mouth, the liquid sweets are projected back into the pharynx, and thence into the throat or œsophagus.

Fig. 10.—Worker extracting nectar from a blossom.

25. It must be observed also, that the tongue is not only flexible but susceptible of inflation, so as to form a sort of bag,* in which

* Dr. Bevan on the Honey Bee, p. 298.

11

the nectar is collected preparatory to being transferred to the œsophagus.

26. The first stomach or honey-bag into which the nectar

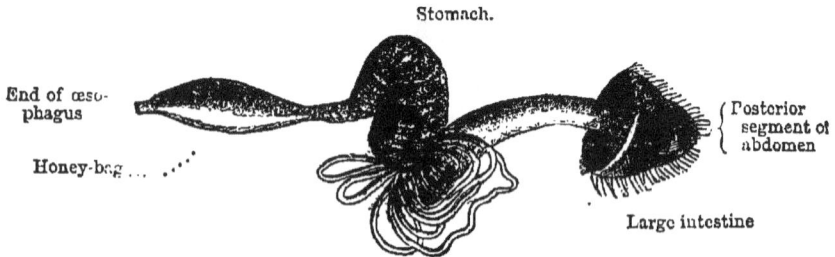

Stomach.

End of œso-
phagus

{ Posterior
segment of
abdomen

Honey-bag ...

Large intestine

Fig. 11.—Digestive apparatus of the Bee (magnified).

passes through the œsophagus,—which is a long and slender tube passing from the back of the mouth through the neck,—has the form of a Florence flask, and is composed of a material as transparent as glass. When filled it has the magnitude of a small pea. The honey received by it is partly regurgitated and deposited for general use in the cells of the comb, which will presently be described. The remainder which constitutes the food of the insect passes into the true stomach, and from thence into the intestines where it undergoes the process of digestion, the products of which are distributed through suitable tubes to all parts of the body for its nourishment.

27. Both the honey-bag and the stomach are susceptible of contraction, by which the food is thrown back from the former into the mouth as in ruminating animals, and from the latter into the intestines.

28. The antennæ are organs of great importance, upon the functions of which, however, naturalists are not fully agreed. It appears certain nevertheless, that they are not only tactile instruments of great sensitiveness, but are organs, by the signs, gestures, and mutual contact of which the bees communicate to each other their mutual wants, and convey information in many cases, some of which will be noticed hereafter, respecting the condition of the hive.

29. The flying-apparatus of the bee, as well as that of many other insects, far exceeds in power the instruments of flight with which the swiftest birds are furnished. To the anterior margin of the under wings are attached eighteen or twenty hooks, which when spread for flight (figs. 5, 6) lay hold of the posterior edges of the upper wings, so that the two wings on each side thus united act as a single wing.

12

30. The three pairs of legs are composed of several joints (fig. 1) articulated like those of the human arm, so as to give great mobility to the member. The lower joints of the two under pairs form brushes, the hairs of which are stiff and bristly, and set upon their inner surfaces. The farina which they collect from the stamina of flowers is swept off by these brushes, as well as by the hairs with which their abdomen and thorax are covered. This farina is afterwards by means of the maxillæ or jaws, and the feet of the anterior pair of legs, rolled into pellets and packed in a pair of spoon-shaped cavities or baskets, provided for that purpose and attached to the feet of the hindmost pair of legs. In this process the brushes, after disposing of their own collection of farina, sweep that flour also from the surface of the abdomen and thorax, and pack it in like manner in the baskets. The exterior of these baskets is smooth and glossy, and the interior lined with strong close hairs to retain the load in its place, and prevent its escape in flight.

Basket

Fig. 12.—Posterior leg of a worker.

It is worthy of remark that neither the queen nor the drones are supplied with this appendage. Since neither exercise any industry they would have no use for it.

31. Each foot terminates in two hooks, the points of which are opposed one to the other. By means of these the insects suspend themselves at will to the sides and roofs of their habitation, and hanging from each other form a living curtain in certain operations which will be presently noticed.

In the middle of each of these is placed the *sucker*, by which the insect is enabled to walk with facility on surfaces with its body downwards, as we see flies walk on ceilings. These suckers are little flexible cups, the edges of which are serrated so as to allow of their close application to any kind of surface. When closely applied, the air between the sucker and the surface is excluded, so that the body is attached to the surface by the pressure of the atmosphere. When the foot is to be detached from the surface, as in walking, the air is readmitted. This apparatus may be

13

easily seen, and its action observed, by inspecting with a microscope the feet of a fly walking on a pane of glass, the observer being on the side of the pane opposite to that on which the fly moves.

32. Besides the stomach and intestines, the abdomen of the queen and workers contains the sting and the apparatus connected with it, by which the venom which it pours into the wound is secreted, an instrument of offence supplied to these in common with many other species of four-winged insects. This formidable weapon of vengeance is established in its tail. All the insects which in common with the bee are supplied with a sting, belong to the order hymenoptera or membrane-winged. This weapon consists of two darts finer than a hair, which lie in juxtaposition, being barbed on the outer sides, but so minutely that the points can only be seen with the microscope. These darts move in the groove of a strong sheath, which is often mistaken for the sting itself. When the dart enters the flesh, a drop of subtle venom, secreted by a peculiar gland, is ejected through the sheath and deposited in the wound. This poison produces considerable tumefaction, attended with very acute pain.

The posterior extremity of the body of a worker with the sting protruded is shown in fig. 13.

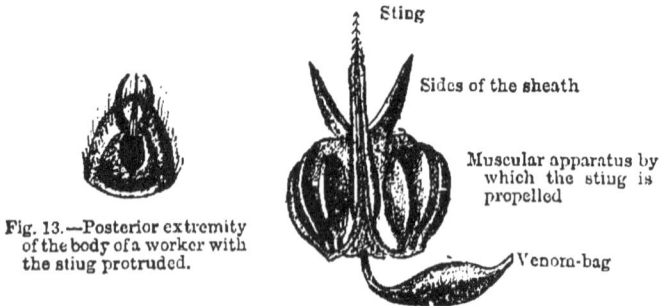

Fig. 13.—Posterior extremity of the body of a worker with the sting protruded.

Fig. 14.—The same slightly magnified, showing the venom-bag.

The sheath of the sting, also called the ovipositor, consists, according to Dr. Bevan, of a long tube, or rather of several tubes, which pass one into another like those of a telescope. The muscles by which the sting is propelled, though too minute to be seen without the microscope, have, nevertheless, sufficient power to drive the sting to the depth of the twelfth of an inch into the thick cuticle of a man's hand. The sting is articulated by thirteen scales to the posterior extremity of the body, and at its root are the pair of glands, one of which appears in fig. 14, in which the poison

14

is secreted. These glands, communicating by a common duct with the groove formed by the junction of the lower parts of the barbed sting, send the venomous liquid through that groove into the wound. On each dart there are four barbs. When the insect intends to sting, one of these piercers having its point a little longer, or more in advance than the others, is first darted into the flesh, and being fixed there by its barb, the other strikes in also; and they alternately penetrate deeper and deeper, till they acquire a firm hold of the flesh with the barbed hooks, and then follows the sheath, enclosing and conveying the poison into the wound. The action of the sting thus, as Paley observed, affords an example of the union of chemical and mechanical principles: of chemistry, in respect to the venom; and of mechanism, in the motion into the flesh. The machinery would have been comparatively useless, had it not been for the chemical process by which in the body of the insect honey is converted into poison; and, on the other hand, the poison would have been ineffectual without an instrument to wound, and a syringe to inject it.

In consequence of the barbed form of the sting, and the strong hold it takes on the flesh, the bee can seldom withdraw it, and in detaching herself from the part stung she generally leaves behind her not only the sting itself, but the venom-bag and a part of her intestines. Swammerdam mentions a case in which even the stomach of the insect was torn from the abdomen in detaching herself, so that in most cases her life is the sacrifice for the gratification of her vengeance.

Although the bee, except in certain cases to be mentioned hereafter, uses its sting only in defence, or for vengeance, when molested, it is sometimes found that it manifests an antipathy to particular individuals, whom it attacks and wounds without provocation.

33. The organs of fecundation and reproduction are also contained in the abdomen. Those of the drone are represented on a magnified scale in fig. 15. They correspond in their functions to those of the superior animals.

Fig. 15.—Apparatus of fecundation of the drone.

The organs of reproduction of the queen, which are objects of considerable interest, are shown on a magnified scale in fig. 16.

34. We have already stated that the king-consort never survives the bridal day. As this does not affect the conjugal fidelity

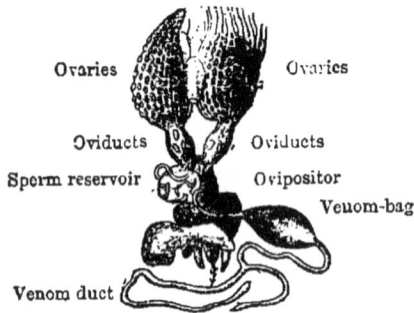

Fig. 16.—Ovaries of the queen and their appendages.

of her majesty, who never allows a successor to her departed lord, so neither does it impose any limit to the posthumous offspring which she bears to him. Small as are the ovaries, or egg organs, which are shown highly magnified in fig. 16, her majesty, according to Huber, generally produces from them about 12000 eggs in the short interval of two months, being at the average rate of 200 per day.

Although her majesty does not continue so prolific during the remainder of her life, she nevertheless gives birth to a progeny enormous in number. The number of eggs deposited by her in the cells in the months of April and May is, as above stated, about 12000. According to Schirach, a prolific queen will lay in a season—that is, from April to October inclusive—from 70000 to 100000 eggs. This amazing power of reproduction is not exerted uniformly during the season. There are two fits, so to speak, of fruitfulness. The first in April and May; the second, in August and September, with an interval of comparative repose in July. This immense increase of population, rendering emigration indispensable, the over-peopled hive sends forth swarm after swarm so fast as the young arrive at maturity; and with each swarm one of the princesses goes forth, and is elevated to the throne of the new colony, except in the event of the abdication of the queen-mother, in which case she emigrates herself, resigning the sovereignty of the hive to one or other of the princesses.

Fig. 76.—Hiving a swarm.

THE BEE.

ITS CHARACTER AND MANNERS.

———◆———

CHAPTER II.

35. This fecundity not anomalous.—36. Bee architecture.—37. Social condition of a people indicated by their buildings.—38. This test applied to the bee. — 39. Individual and collective habits. — 40. Solitary bees.—41. Structure of their nests.—42. Situation of nests. —43. Anthidium manicatum. —44. Expedient for keeping nest warm.—45. Clothier bee.—46. Carpenter bee.—47. Mason bee.—48. Expedient to protect the nest.—49. Upholsterer bee.—50. Hangings and carpets of her rooms.—51. Leaf-cutter bees.—52. Method of making their nest.—53. Process of cutting the leaves.—54. Hive-bee.—55. Structure of the comb.—56. Double layer of cells.—57. Pyramidal bases.—58. Illustrative figures.—59. Single cells.—60. Combination of cells.—61. Great advantages of hexagonal form.—62. Economy of space and material.—63. Solidity of structure.—64. Geometrical problem of the comb solved.—65. Expedient to secure the sides and bases of the cells.

35. The prodigious fecundity of the queen of the bees is by no means an anomaly in the insect world. The female of the white ants produces eggs at the rate of one per second, or 3600 per hour, or 86400 per day. Now, although this insect certainly does not

lay at this rate all the year round, yet, taking the lowest estimate of the period of her reproduction, the number of her young will probably exceed not only that of the queen bee, but that of any other known animal.*

36. There is nothing in the economy of the bee more truly wonderful, nor more calculated to excite our profound veneration of the beneficent power, which conferred upon it the faculties which guide its conduct, than the measures which it takes for the construction of its dwelling, and for those of its young. These processes are very various, according to the particular species of the insect which executes them. Now, most of these species differ in the mechanical and architectural principles upon which they base the construction of their dwellings, all agreeing, nevertheless, in this, that they select those principles with admirable skill, adapting them in all cases to the situation and circumstances in which their habitations are erected.

37. If we would form an estimate of the civilisation and intellectual condition of the population of a newly-discovered country, we usually direct our attention, as Kirby observes, to their buildings and other examples of architectural skill. If we find them like the wretched inhabitants of Van Diemen's land, without other abodes than natural caverns, or miserable penthouses of bark, we at once regard them as ignorant and unhumanised. If, like the South Sea islanders, they live in houses of timber thatched with leaves, and supplied with various utensils, we place them much higher in the scale. But when we discover a nation inhabiting towns like the ancient Mexicans, consisting of stone houses regularly arranged in streets, we do not hesitate to pronounce them advanced to a considerable point in civilisation.

If, moreover, it be found that each building has been constructed upon the most profound mathematical principles, so that the materials have been applied under such conditions as ensure the greatest degree of strength, combined with the greatest degree of lightness; and that, while the internal apartments display the most beautiful symmetry, they also afford the greatest capacity which a given amount of materials can admit, we at once arrive at the conclusion that such a population must have arrived not alone at the highest degree of civilisation, but at the highest point in the advancement of the sciences.

38. If we were to affirm that all this may be said with the most rigorous truth of many varieties of the bee, and above all of the common hive-bee, we might be suspected of being merely excited by that enthusiasm so common with those, who devote

* See Tract on the White Ants.

themselves exclusively to one particular pursuit. We must, nevertheless, leave the reader to judge how far such a statement is chargeable with the exaggeration of enthusiasm, when he shall have duly pondered upon all that we shall explain to him in the following pages; and if, perchance, his wonder be raised to the point of incredulity, that sentiment will be repressed when he remembers, *who taught the bee !*

39. Bees, like the human race, sometimes exercise their industry individually and sometimes collectively. Their habitations also are sometimes constructed exclusively for their young, and may be called *nests* rather than *dwellings*. This is more especially the case with solitary insects. In the case of social bees, which live together in organised communities, the habitations are generally adapted as well for the members of the colony themselves, as for their progeny.

40. The operations of these solitary insects, though exhibiting, as will presently appear, marvellous skill, are infinitely inferior to those of the social bees. We shall, therefore, first notice the more simple labours of the former.

41. Among the most inartificial structures executed by the solitary species, are the habitations of the *colletes succinctæ, fodiens*, &c. The situation chosen in these cases is either a bank of dry earth, or the cavities of mud walls. A cylindrical hole pierced in a horizontal direction about two inches in length is first produced. The bee makes in this three or four thimble-shaped cells, each of which is about a sixth of an inch in diameter and half an inch long, fitting one into another like thimbles. The materials of these cells is a silky membrane resembling gold-beater's leaf, but much finer, and so very thin and transparent that the form and colour of any enclosed object can be seen through it. This material is secreted by the insect. When the first of these cells is completed, the insect deposits in it an egg and fills it with a pasty substance, which is a mixture of pollen and honey. When this is done she proceeds to form the second cell, inserting its end in the mouth of the first as above described, and in like manner lays an egg in it and deposits with it a like store of food for the future young. This goes on until the cylindrical hole receives three or four cells which nearly fill it. The bee then carefully stops up the mouth of the hole with earth.

42. The situations in which these simple nests are placed are very various. They are not only found as above stated in banks of earth and mud walls, and the interstices of stone walls, but often also in the branches of trees. Thus a series of them was found by Grew in the pith of an old elder branch.

43. Some varieties of the bee, such as the *anthidium manicatum*, dispense with the labour of boring the cylindrical holes above

described, and avail themselves of the ready-made cavities of trees, or any other object which answers their purpose. Kirby mentions the example of nests of this kind found by himself and others, constructed in the inside of the lock of a garden-gate.

44. A proceeding has been ascertained on the part of these insects in such cases, which it is extremely difficult to ascribe to mere instinct, independent of some intelligence. Wherever the nest may be constructed, the due preservation of the young requires that until they attain the perfect state, their temperature should be maintained at a certain point. So long as the material surrounding their nest is a very imperfect conductor of heat, as earth or the pith of wood is, the heat developed by the insect, being confined, is sufficient to maintain its temperature at the requisite point. But if, perchance, the mother-bee select for her nest any such locality as that of the lock of a gate, the metal, being a good conductor of heat, would speedily dissipate the animal heat developed by the insect, and thus reduce its temperature to a point incompatible with the continuance of its existence. How then does the tender mother, foreseeing this, and consequently informed by some power of the physical quality peculiar to the metal surrounding the nest, provide against it? How, we may ask, would a scientific human architect prevent such an eventuality? He would seek for a suitable material which is a non-conductor of heat and would surround the nest with it. In fact the very thing has occurred in a like case in relation to steam-engine boilers. The economy of fuel there rendered it quite as necessary to confine the heat developed in the furnace, as it is to confine that which is developed in the natural economy of the pupa of the bee. The expedient therefore resorted to is to invest the boiler in a thick coating of a sort of felt, made for the purpose, which is almost a non-conductor of heat. A casing of sawdust is also used in Cornwall for a like purpose. By these expedients the escape of heat from the external surface of the boiler is prevented.

45. The bee keeps its pupa warm by an expedient so exactly similar, that we must suppose that she has been guided either by her own knowledge, or by a power that commands all knowledge, in her operations. She seeks certain woolly leaved plants, such as the stachys lanata or the agrostemma coronaria, and with her mandibles scrapes off the wool. She rolls this into little balls, and carrying it to the nest, sticks it on the external surface by means of a plaster, composed of honey and pollen, with which she previously coats it. Thus invested, the cells become impervious to heat, and consequently all the heat developed by the little animal is confined within them.

20

This curious habit of swathing up its pupa in a kind of warm blanket has given to these species the name of *clothiers.*

46. Another class of bees has acquired the name of *carpenters,* from the manner in which they carve out their nest in wood-work. This bee, which is represented in fig. 17, and of which the nest is shown in fig. 18, having been already described in our Tract on Instinct and Intelligence (72), need not be noticed further here.

Fig. 17.—The Carpenter Bee.
(Xylacope).

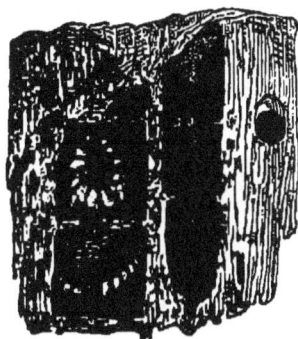

Fig. 18.—Nest of the Carpenter Bee.

47. Another class of this insect has acquired the name of *masons,* from the circumstance of building their nests of a sort of artificial stone. The situation selected is usually a stone wall, having a southern aspect, and sheltered on either side by some angular projection. The situation being decided upon, the mother-bee proceeds to collect the materials for the mansion, which consist of sand, with some mixture of earth. These she glues together, grain by grain, with a cement composed of viscid saliva, which she secretes. Having formed this material into little masses, like the grains of small shot, she transports them with her mandibles to the place where she has laid the foundation of her mansion.

With a number of these masses, united together by an excellent cement secreted by her organs, she first lays the foundation of the building. She next raises the walls of a cell about an inch in length, and half an inch broad, resembling in form a thimble. In this she deposits an egg, fills it with a mixture of pollen and honey, in the same manner as described in the former case, and after carefully covering it in, proceeds to the erection of a second building of the same kind, which she furnishes in the same manner, and so continues until she has completed from four to eight.

These cells are not placed in any regular order; some are

21

parallel, others perpendicular, and others inclined to the wall at different angles. The whole mass is consolidated by filling up the irregular intersticial spaces between the cells, with the same material as that of which the walls are built. After this has been accomplished, the whole is covered up with coarser grains of sand.

The nest when thus finished resembles a mass of solid stone, so hard as to be cut with much difficulty by a knife. Its form is an irregular oblong, and to a casual observer presents the appearance of a mere splash of mud rather than that of a regular structure.

The insects are sometimes so sparing of their labour, that they avail themselves of old nests when they can find them, and often have desperate combats to seize and retain possession of them.

48. It might be imagined that nests so solidly constructed would afford perfect protection to the young from its enemies; such is nevertheless not found to be the case. The ichneumon and the beetle both contrive occasionally to deposit their eggs in the cells, the larvæ of which never fail to devour their inhabitants.

Different varieties of the masons select different situations and materials for their nests. Some use fine earth, which they make into mortar with gluten. Others mix sandy earth with chalk. Some construct their nests in chalk-pits, others in the cavities ot large stones, while others bore holes for them in rotten wood. Wherever placed they endeavour to conceal them, by plastering or covering them with some material different from that of which the nest is constructed. Thus one species surrounds its nest with oak-leaves glued to its surface. M. Goureau mentions the case of a bee that employed an entire day, in arranging blades of grass about two inches long, in the form of the top of a tent over the mouth of its nest. A case of this sort was also observed by Mr. Thwaites, who saw a female for a considerable time collecting small blades of grass, which she laid over the empty shell of a snail in which she had located her nest.

49. The name of *upholsterers* has been given by Kirby to certain species of bees, who, having excavated their nest in the earth, hang its walls with a splendid coating of flowers and leaves. One of the most interesting of these varieties is the *megachile papaveris*, which has been described by Reaumur. It chooses invariably for the hangings of its apartments the most brilliant scarlet, selecting as its material the petals of the wild poppy, which the insect dexterously cuts into the proper form.

50. Her first process is to excavate in some pathway a burrow cylindrical at the entrance, but enlarged as it descends, the depth being about three inches. After having polished the walls, she next flies to a neighbouring field, where she cuts out the oval

parts of the poppy blossoms, and seizing them between her hind legs returns with them to her cell. Sometimes it happens that the flower from which she cuts these, being but half blown, has a wrinkled petal. In that case she spreads out the folds, and smoothes away the wrinkles, and if she finds that the pieces are too large to fit the vacant spaces on the walls of her little room, she soon reduces them to suitable dimensions, by cutting off all the superfluous parts with her mandibles. In hanging the walls with this brilliant tapestry she begins at the bottom, and gradually ascends to the roof. She carpets in the same manner the surface of the ground round the margin of the orifice. The floor is rendered warm sometimes by three or four layers of carpeting, but never has less than two.

Our little upholsterer having thus completed the hangings of her apartment, fills it with a mixture of pollen and honey to the height of about half an inch. She then lays an egg in it, and wraps over the poppy lining, so that even the roof may be furnished with this material. Having accomplished this she closes the mouth of the nest.*

51. It is not every insect of this class which manifests the same showy taste in the colours of their furniture. The species called *leaf-cutters* hang their walls in the same way, not with the blossoms but the leaves of trees, and more particularly those of the rose-tree. They differ also from the upholsterer, described above, in the external structure of their nests, which are formed in much longer cylindrical holes, and consist of a series of thimble-shaped cells, composed of leaves most curiously convoluted. We are indebted likewise to Reaumur for a description of the labours of these.

52. The mother first excavates a cylindrical hole in a horizontal direction eight or ten inches long, either in the ground or in the trunk of a rotten tree, or any other decaying wood. She fills this hole with six or seven thimble-shaped cells, composed of cut leaves, the convex end of each fitting into the open end of the other. Her first process is to form the external coating, which is composed of three or four pieces of larger dimensions than the rest, and of an oval form. The second coating consists of portions of equal size, narrow at one end, but gradually widening towards the other, where the width equals half the length. One side f these pieces is the serrated edge of the leaf from which it was taken, which, as the pieces lap over each other, is kept on the outside, the edge which was cut being within.

The little animal next forms a third coating of similar material,

* Reaumur, vi. 139 to 148.

the middle of which, as the most skilful workman would do in a like case, she places over the margins of those that form the first side, thus covering and strengthening the junctions by the expedient which mechanics call a break-joint. Continuing the same process she gives a fourth and sometimes a fifth coating to her nest, taking care at the closed end or narrow extremity of the cell, to bend the leaves so as to form a convex termination.

After thus completing each cell, she proceeds to fill it to within the twentieth of an inch of the orifice with a rose-coloured sweet-meat made of the pollen collected from thistle blossoms mixed with honey. Upon this she lays her egg, and then closes the orifice with three pieces of leaf, one placed upon the other, concentrical and also so exactly circular in form, that no compasses could describe that geometrical figure with more precision. In their magnitude also they correspond with the walls of the cell with such a degree of precision, that they are retained in their situation merely by the nicety of their adaptation.

The covering of the cell thus adapted to it being concave, corresponds exactly with the convex end of the cell which is to succeed it, and in this manner the little insect prosecutes her maternal labours, until she has constructed all the cells, six or seven in number, necessary to fill the cylindrical hole.

53. The process which one of these bees employs in cutting the pieces of leaf that compose her nest, is worthy of attention. Nothing can be more expeditious, and she is not longer about it than one would be in cutting similar pieces with a pair of scissors. After hovering for some moments over a rose-bush, as it were to reconnoitre the ground, the bee alights upon the leaf which she has selected, usually taking her station upon its edge, so that its margin shall pass between her legs. She then cuts with her mandibles, without intermission, in such a direction as to detach from the leaf a triangular piece. When this hangs by the last fibre, lest its weight should carry her to the ground, she spreads her little wings for flight, and the very moment the connection of the part thus cut off with the leaf is broken, she carries it off in triumph to her nest, the detached portion remaining bent between her legs in a direction perpendicular to her body. Thus, without rule or compass, do these little creatures measure out the material of their work into ovals, or circles, or other pieces of suitable shapes, accurately accommodating the dimensions of the several pieces of these figures to each other. What other architect could carry impressed upon the tablet of his memory such details of the edifice which he has to erect, and destitute of square or plumb-line, cut out his materials in their exact dimensions without making a single mistake or requiring a single subsequent correc-

tion ?´ Yet this is what the little bee invariably does. So far are human art and reason surpassed by that instruction which the insect receives from its Divine Creator.*

54. But of all the varieties of this insect, that of which the architectural and mechanical skill is transcendently the most admirable, is the *hive-bee*. The most profound philosopher, says Kirby, equally with the most incurious of mortals, is filled with astonishment at the view of the interior of a bee-hive. He beholds there a miniature city. He sees regular streets, disposed in parallel directions, and consisting of houses constructed upon the most exact geometrical principles, and of the most symmetrical forms. These buildings are appropriated to various purposes. Some are warehouses in which provisions are stored in enormous quantities. Some are the dwellings of the citizens, and a few of the most spacious and magnificent are royal palaces. He finds that the material of which this city is built, is one which man with all his skill and science cannot fabricate, and that the edifices which it is employed to form are such that the most consummate engineer could not reproduce, much less originate; and yet this wondrous production of art and skill is the result of the labour of a society of insects so minute, that hundreds of thousands of them do not contain as much ponderable matter, as would enter into the composition of the body of a man. *Quel abîme aux yeux du sage qu'une ruche d'abeilles ! Quelle sagesse profonde se cache dans cet abîme ! Quel philosophe osera le sonder !* Nor has the problem thus solved by the bee, yet been satisfactorily expounded by philosophers. Its mysteries have not yet been fathomed. In all ages naturalists and mathematicians have been engrossed by it, from Aristomachus of Soli and Philiscus the Thracian, already mentioned, to Swammerdam, Reaumur, Hunter, and Huber of modern times. Nevertheless the honey-comb is still a miracle which overwhelms our faculties.†

55. A honey-comb, when examined, is found to be a flattish cake with surfaces sensibly parallel, each surface being reticulated with hexagonal forms of the utmost regularity. No geometrician could describe the regular hexagon with greater precision than is here exhibited.

It is proved in geometry that there are only three regular figures, which, being joined together at their corners, will so fit each other as to leave no unoccupied spaces between them. These figures are the square, the equilateral triangle, and the regular hexagon. Four squares united by one of their angles will fill all

---

* Reaumur, vi. 971 ; Kirby, Int., i. 377.
† Kirby, i. 410.

the surrounding space, and any number of squares may thus be combined so as to cover a surface like a mosaic pavement without leaving any intermediate unoccupied spaces.

In like manner six equilateral triangles will have a like property, and in fine, three regular hexagons being similarly united at one of their corners, will in like manner completely occupy the surrounding space.

Since no other regular geometrical figure possesses this property, it follows that a regular mosaic pavement must necessarily be composed of one or other of these figures.

Fig. 19 represents such a pavement composed of squares; and fig. 20, one composed of equilateral triangles; and in fine, fig. 21, one composed of regular hexagons.

Fig. 19.

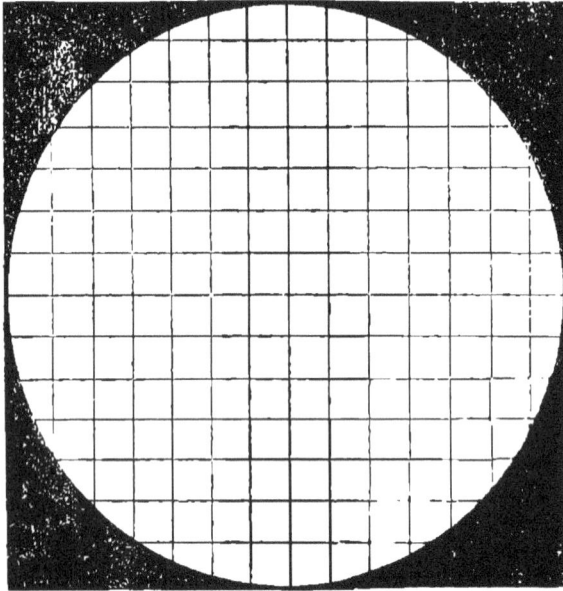

The angles, in fig. 19, are 90°; those in fig. 20, are 60°; and those in fig. 21, 120°. No other angles save these, therefore, could be used in any regular pavement of this kind without leaving intersticial uncovered spaces.

Now it will be at once perceived that the form presented by the surface of a honey-comb is that of an hexagonal pavement. We shall presently see why the bee has selected this in preference to either of the other possible forms.

26

56. On further examining the comb, it will be found that the hexagonal spaces presented by its surface are the mouths of so

Fig. 20.

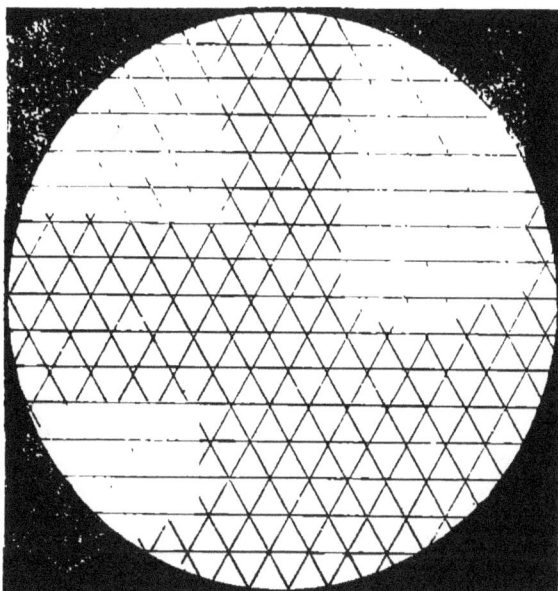

many hexagonal tubes which are filled with honey. If any of these be empty, it will be seen that the depth of these tubes is half the thickness of the comb.

57. It appears therefore that the honey-comb is a combination of hexagonal tubes, placed in juxtaposition, the angles of the hexagon being fitted into each other like the stones of a mosaic pavement; that there are two systems of such tubes, meeting in the middle of the thickness of the comb, their mouths being presented outwards on both sides, and consequently their bases resting against each other.

If by the dissection of the comb, the forms of their bases be examined, they will be found to consist, not as might be at first supposed of plane regular hexagons, which would be the case if they were plane surfaces at right angles to the tube; they will be found, on the other hand, to have the form of pyramids, each of which is composed of three regular lozenges united together at their edges, so as to form an apex; this apex being pointed always towards the opposite side of the comb. The pyramidal base is

27

thus a geometrical figure, having as much regularity as the
hexagonal tube, of which it forms the termination, but constructed

Fig. 21.

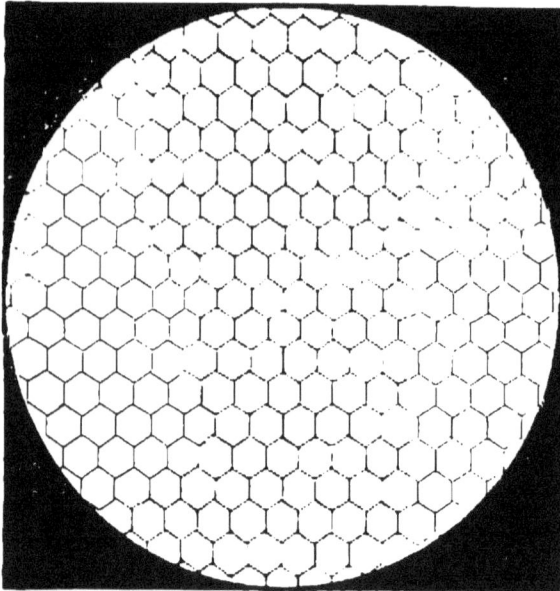

on a totally different principle. The angles of the lozenges,
which form its sides, are one obtuse and the other acute ; and these
pyramidal bases of the cells, on one side of the comb, fit into
corresponding cavities, made by the similar pyramidal bases of the
cells, on the other side of the comb, so as to leave no intermediate
unoccupied space.

58. Without the aid of perspective figures, and even with such
aid, without some effort of imagination on the part of the reader,
it would be impossible to convey a clear notion of this part of the
structure of the honey-comb, and yet without such a clear notion
it would be totally impossible to appreciate the admirable results
of bee industry. We have, therefore, attempted to represent in
figs. 22 and 23, the bases of four contiguous cells seen from the
inside and from the outside. In fig. 22 is presented an inside view
of the bases of three adjacent cells, *a a a*. It must be observed that
a a a are here intended to represent angular cavities, each formed
by the junction of three lozenge-shaped planes, such as have been
just described. Now it will be seen, that as a necessary consequence
of this juxtaposition, a figure will be formed at *b*, by three lozenge-

28

shaped planes, one belonging to each of the three bases, *a a a*, and that this, instead of being hollow on the side presented to

Fig. 22. Fig. 23. Fig. 24. Fig. 25.

 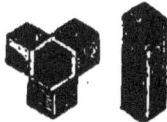 

the eye, will be hollow on the opposite side, which is turned from the eye, and will there form an angular cavity precisely similar and equal to the cavities *a a a*, which are turned towards the eye. Now this cavity, which is thus turned to the opposite side, is the base of one of the cells on the other side of the comb. In fig. 23 we have presented a view of the combination as it would be seen on the other side. In this case, the angular cavity darkly shaded in the middle of the figure, is the angular projection, *b*, in fig. 22, seen on the other side; and the three angular projections which surround it, jutting forward towards the eye, are the three angular bases, *a a a*, fig. 22, seen on the other side.

59. A perspective view of a single hexagonal tube or cell, with its pyramidal base, is shown in fig. 24.

The manner in which the hexagonal cells are united base to base to form the comb, is shown in perspective in fig. 25, where *a* is the open mouth of the tube, and *b c* the lozenge-shaped planes, forming the bases of the opposite tubes. The same is shown in section in fig. 26.

Fig. 26. Fig. 27. Fig. 28.

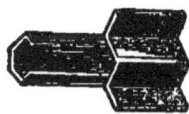  

60. Several hexagonal cells are shown in their natural juxtaposition, placed base to base, as they form the comb, in fig. 27, and a perspective view of their pyramidal bases is given in fig. 28.

Nothing can be more surprising than this production of such an insect, when regarded as a piece of scientific engineering. The substance which comprises it being one secreted by the bees in limited quantity, it was of the greatest importance in its use, that a material so scarce should be applied so as to produce the greatest possible result, with the smallest possible quantity of the material. The problem, therefore, which the bee had to solve

29

was, with a given quantity of wax, to construct a combination of similar and equal cells of the greatest aggregate capacity, and such as to occupy the available space in the hive to the greatest possible advantage. The form and magnitude of the cells must necessarily have been adapted to those of the bee itself, because these cells are intended to be the nests in which the eggs are laid and hatched, and the young bee raised to its state of maturity.

The body of the bee being oblong, and measuring about six-tenths of an inch in length by two-tenths in diameter, cylindrical tubes of corresponding dimensions would have answered the purpose; but such tubes could not be united together in juxtaposition without either a great waste of wax or great deficiency of strength, since, when placed in contiguity, they would leave between them empty spaces of considerable magnitude, which, if left unoccupied, would render the structure weak, and if filled with wax, would have the double disadvantage of giving needless and injurious weight to the comb, and involving the waste of a quantity of a scarce and precious material, greater than all that would be necessary to form the really useful part of the comb.

61. From what has been explained it will be understood that, to form a combination of tubular cells without interstices, the choice of the bee was necessarily limited to the three figures already mentioned—the equilateral triangle, the square, and the regular hexagon. The equilateral triangle would be attended with the disadvantage of a great waste of both space and material; for if its dimensions were sufficient to afford easy room to the body of the bee, a large space would be wasted at each of the angles, towards which the body of the bee could never approach.

A like disadvantage, though less in degree, would have attended square tubes. The bee, therefore, with the instinct of an engineer, decided on the third form, of the regular hexagon, which at once fulfilled the conditions of a sufficiently near adaptation to the form of its own body, and the advantage of such a combination as would leave neither waste space nor loss of material.

62. In the structure of the comb there is still another point worthy of attention. It might naturally have been expected that it would be composed of a single layer of cells, one side presenting the mouth, and the other the pyramidal base; but if this had been the course adopted, the side consisting of the pyramidal bases would be an extensive surface, upon which the industry of the bee would have no occupation, and the space in the hive to which such surface would be presented would, therefore, be so much space wasted. Instead, therefore, of constructing the comb of a single layer of cells, the bees judiciously make it of a

double layer, the pyramidal bases of each layer being placed in contact with each other.

It might also have been expected that these bases would have received the most simple form of plane surfaces, so that the side of each layer occupied by them would be a uniform plane ; and these planes resting in contact would form the comb ; but to this there would be several objections. In the first place, the capacity of the comb would be less ; the bases of the cells, placed in contact, would be liable to slip one upon the other ; and if the cells had a common base, they would have less strength ; but independently of this, the bee itself tapers towards its posterior extremity, and a cell with a flat bottom having no corresponding tapering form would be little adapted to its shape, and would involve a consequent waste of space. The bee avoids this disadvantage by giving the bottom of the cell the shape of a hollow angular pyramid, into the depth of which the tapering posterior extremity of the insect enters.

63. There is another advantage in this arrangement which must not be overlooked. The pyramidal bases of each layer of cells, placed in juxtaposition by reciprocally fitting each other, so that the angular projections of each are received into the angular cavities of the other, are effective means of resisting all lateral displacement.

64. Pyramidal bases, however, might have been given to the cells in a great variety of ways, which would have equally served the purposes here indicated ; but it was essential, on grounds of economy, that that form should be selected which would give the greatest possible capacity with the least possible material. On examining curiously the form of the lozenges composing the pyramidal bases of the cells, Maraldi found by accurate measurement that their acute angle measured 70° 32', and consequently their obtuse angle 109° 28'. Magnitudes so singular as these, invariably reproduced in all the regular cells, could scarcely be imagined to have been adopted by these little engineers without a special purpose, and Reaumur accordingly conjectured that the object must have been the economy of wax.

Not being himself a mathematician sufficiently profound to solve a problem of this order, he submitted to M. Koenig, an eminent geometer of that day, the general problem to determine the form which ought to be given to the pyramidal bottom of an hexagonal prism, such as those constituting the cones, so that with a given capacity, the least possible material would be necessary for the construction. The problem was one requiring for its solution the highest resources to which analytical science had then attained. Its solution, however, was obtained, from which it appeared that the proper angles for the lozenges would be 70° 34′ for the acute, and consequently 109° 26′ for the obtuse angle. Here are then in juxtaposition the result of the labours of the geometer and the bee.

	ACUTE ANGLE.	OBTUSE ANGLE.
Geometer	70° 34′	109° 26′
Bee	70° 32′	109° 28′

We leave the reader to enjoy the contemplation of these numbers without one word more of comment.

65. " Besides the saving of wax effected by the form of the cells, the bees adopt another economical plan suited to the same end. They compose the bottoms and sides of wax of very great tenuity, not thicker than a sheet of writing-paper ; but as walls of this thickness at the entrance would be perpetually injured by the ingress and egress of the workers, they prudently make the margin at the opening of each cell three or four times thicker than the walls. Dr. Barclay discovered that though of such excessive tenuity, the sides and bottom of each cell are actually double, or in other words, that each cell is distinct, separate, and in some measure an independent structure, agglutinated only to the neighbouring cells ; and that when the agglutinating substance is destroyed, each cell may be entirely separated from the rest. This, however, has been denied by Mr. Waterhouse, and seems inconsistent with the account given by Huber, hereafter detailed; but Mr. G. Newport asserts, that even the virgin-cells are lined with a delicate membrane." *

* Kirby, i. p. 412.

Fig. 55.—COVERED APIARY.

THE BEE.

ITS CHARACTER AND MANNERS.

—•—

CHAPTER III.

66. Drone cells and worker cells.—67. Store cells.—68. Construction of combs.—69. Wax-makers also produce honey.—70. First operation of the wax-makers.—71. Process of the foundress.—72. Kneading the wax.—73. Formation of first wall.—74. Correction of mistakes.—75. Dimensions of first wall.—76. Operations of the nurses.—77. Bases of cells.—78. Wax-makers resume their work. —Completion of pyramidal bases.—79. Pyramidal partition.—80. Formation of cells.—81-82. Arrangement of combs.—83. Sides not parallel.—84. Process not merely mechanical.—85-86. Process of construction.—87. Labour successive.—88. Dimensions of cells.—89. Their number.—90. Bee-bread.—91. Pap for young.—92. Food adapted to age.—93. Transformation.—94. Humble-bees—females. —95. Their nursing workers.—96. Transformation.—97. How the temperature of the cocoons is maintained.—98. Anecdote related by Huber.—99. Remarkable care of the nurses.—100. Heat evolved in respiration by the hive-bee—101. Cross alleys connecting the streets. —102. First laying of the queen in Spring.—103. Her royal suite.—104. The eggs.

66. Since the population of the hive is composed, as already explained, of different classes of individuals having different stature, and since one of the purposes of the cells is to be their

abode from the time they issue from the egg until they attain maturity, it follows that the capacity of the cells, or such of them as are thus appropriated, must be subject to a corresponding difference. The cells of the workers will therefore be less in magnitude than those of the drones, and these last much less than the royal cells. The comb therefore consists of different parts reticulated by hexagons of different magnitudes, the smaller ones being the mouths of the cells appropriated to the workers, and the larger those of the cradles of the drones. As to the royal cells they differ altogether from the others, not only in capacity, but also in position and form. As already explained, the general forms of the cells are hexagonal tubes, with pyramidal bases, and open mouths ranged horizontally, their axes being at right angles to the flat sides of the comb. The comb itself is placed vertically in the hive, and the royal cells which are large and pear-shaped are cemented to its lower edges, hanging from it vertically like stalactites from the roof of a cavern. Although there be but one queen in each hive, she produces, nevertheless, three or four or more, and sometimes even as many as thirty or forty royal eggs. The princesses which issue from these, are destined to be the queens of the successive swarms which the hive sends forth.

67. The cells which are appropriated exclusively to the storage of honey and pollen, are similar in form and position to those appropriated to the young drones and workers, but are greater in length, and this length the bees vary according to the exigencies of their store of provisions. If more of these result from their labours than the cells constructed can contain, and there is not time or space for the construction of more cells, they lengthen the honey-cells already made by cementing a rim upon them. They sometimes also use for storage, cells which have already been occupied by young drones or workers, which, having attained their state of maturity, have vacated them.

68. Having thus explained in general the forms and structures of the cells, we shall briefly explain the operation by which the bees construct them, and by their combination form the combs.

The material of the combs is *wax*, a substance secreted beneath the ventral segments of the bodies of that class of the workers which, from this circumstance, has received the name of wax-makers. The apparatus by which the material which ultimately acquires the character of wax is secreted, consists of four pairs of membranous bags, called wax-pockets, which are situated at the base of each segment of the body, one on each side, and which in the natural condition of the body, are concealed by the segments overlapping each other. They can, however, be rendered visible by drawing out the body longitudinally, so that the part

34

of each segment covered by the preceding one shall be disclose (fig. 29).

In these pockets the substance to be ultimately converted into wax is secreted from the food taken into the stomach, which, transpiring from thence through the membrane of the wax-pocket, is formed there in thin laminæ. The stomach and its appendages which are endowed with these functions, though much less capacious in the nurses than in the wax-makers, is not altogether absent; and the nurses have a certain limited power of secreting wax. In them the wax-making function, however, seems to exist in little more than a rudimentary state.

Fig. 29.

69. Although the chief duty of the wax-makers is that from which they have taken their names, they are also capable of producing honey, and when the hive is abundantly furnished with combs, they accordingly change the object of their industry and produce honey instead of wax.

70. When a comb is about to be constructed, the operation is commenced by the wax-makers, who, having taken a due portion of honey or sugar, from either of which wax can be elaborated,

Fig. 30.

suspend themselves one to another—the claws of the fore-legs of the lowermost being attached to those of the hind-legs of the next above them, so that they form a cluster, the external surface of which presents the appearance of a fringed curtain (fig. 30). After having remained in this state unmoved for about twenty-four hours, during which period the material of the wax is secreted, the thin laminæ into which it is formed may generally be perceived under the abdomen.

A single bee is now seen to separate itself from the cluster and to pass from among its companions to the roof of the hive, where by turning itself round, it clears a circular space for its work, about an inch in diameter. Having done this, it proceeds to lay the foundation of a comb in the following manner, if one may be permitted to apply the word foundation to the top of a suspended structure.

71. The foundress bee, as this individual is called, commences its work by seizing with one of its hind feet a plate of wax, or rather of the material out of which wax is to be constituted, from between the segments of its abdomen. The insect is

represented in this act in fig. 31. Having fixed a secure hold on the lamina, it carries it by its feet from the abdomen to its mouth, where it is taken by one of the fore-legs which holds it vertically while the tongue rolled up serves for a support, and by raising and depressing at will, causes the whole circumference to be brought successively under the action of the mandibles (fig. 32), so that the margin is soon ground into pieces. These pieces fall gradually as they are detached in the double cavity of the mandibles which are bordered with hair.

Fig. 31. Fig. 32.

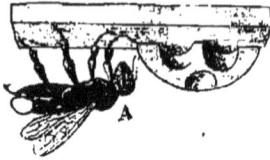 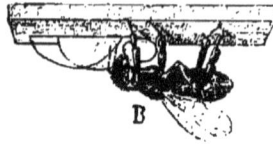

The mandibles or jaws which execute this process open in a horizontal, instead of a vertical, direction as in the case of the superior animals, and have a form resembling that of a pair of shears or scissors.

72. The fragments of the laminæ thus divided falling on either side of the mouth, and pressed together into a compact mass, issue from it in the form of a very narrow ribbon. This ribbon is then presented to the tongue by which it is impregnated with a frothy liquor, which has the same effect upon it as water has on potter's earth in the formation of porcelain paste. That this process, by which the raw material of the wax is worked and kneaded, is an extremely elaborate and artificial one, is rendered apparent by observing carefully the manœuvres of the bee's tongue in the process. Sometimes that organ assumes the form of a spatula, or apothecary's knife, sometimes it takes the form of a mason's trowel, and sometimes that of a pencil tapering to a point, never ceasing to work upon the ribbon which is being evolved from the mouth in these several forms.

After the ribbon has been thus thoroughly impregnated with moisture, and carefully kneaded, the tongue again pushes it between the mandibles, but in a contrary direction to that in which it previously passed, when the whole is worked up anew.

The substance is now converted into true wax, the characteristic properties of which it has acquired in this process. The material evolved in laminæ from the segments of the abdomen is brittle and friable, and would be as unfit for the structure of the comb as dry potter's earth would be for the formation of a vase. The liquid secreted from the mouth, with which it has been impreg-

nated, and the elaborate process of kneading which it has undergone, have totally changed its mechanical properties and have
imparted to it that ductility and plasticity so eminently characteristic of wax. It has also undergone other physical changes.
The laminæ taken from the abdominal segments are colourless
and transparent, the substance into which they are converted,
being white and opaque. .

73. The pieces of wax thus elaborated the insect applies against
the roof of the hive, arranging them with her mandibles in the
intended direction of the comb. She continues thus until she
has in this way applied the wax produced from the entire laminæ,
when she takes in like manner another from her abdomen, treat-
ing it in the same way. After thus heaping together all the wax
which her organs have secreted, and causing it to adhere by its
proper tenacity to the vaults of the hive, she withdraws from her
work and is succeeded by another labourer who continues the
same operations, who is followed in a like manner by a third and
fourth, and so on, all disposing the produce of their labour in the
direction first intended to be given to the comb.

74. Nevertheless it would seem that the curious facility by
which these proceedings are directed is not altogether unerring, for
it happens by chance now and then that one of the workers will
commit a mistake by placing the wax in the wrong direction. In
such cases, the worker which succeeds never fails to rectify the
error, removing the materials which are wrongly placed, and
disposing them in the proper direction.

75. The result of all these operations of the wax-makers is the
construction of a rough wall of wax about half an inch long, a
sixth of an inch high, and the twenty-fourth of an inch thick,
which hangs vertically from the roof of the hive. In the first
rough work there is no angle nor the least indication of the
form of the cells. It is a mere straight and plain vertical partition of wax, roughly made, about the twenty-fourth of an inch
thick, and such as can only be regarded as the foundation of a
comb.

76. The duty of the *wax-makers* terminating here, they are
succeeded by the *nurses*, who are the genuine artisans; standing
in relation to the wax-makers in the same manner as, in the construction of a building, the masons who work up the materials into
the form of the intended structure would to the common labourers.
One of the nurses commences its operation by placing itself horizontally on the roof of the hive, with its head presented to the
wall of wax constructed by the wax-makers. This wall or
partition is intended to be converted into the system of pyramidal
bases of the cells already described, and accordingly the first

37

labour of the nurses is directed to accomplish this change. Their first operation, therefore, is to mould on that side of the wall to which its head is directed, a pyramidal cavity having the form of the base of one of the intended cells. When it has laboured for some minutes thus, it departs and is succeeded by another, who continues the work, deepening the cavity and increasing its lateral margins by heaping up the wax on either side by means of its teeth and fore-feet, so as to give the sides a more regular form. More than twenty nurses succeed each other in this operation.

77. It must be remembered that during this process, nothing has been done on the other side of the partition, but when the cell just described has attained a certain length, other nurses approach the opposite side of the partition and commence the formation of the pyramidal base of two cells corresponding in position with that just described, and these in like manner prosecute their labours, constantly relieving each other.

78. While the nurses are thus employed in converting the rough partition into the pyramidal bases of cells, and in forming the hexagonal tubes corresponding to these pyramidal bases, the wax-makers return and, resuming their labour, increase the magnitude of the partition in every direction, the nurses meanwhile still prosecuting their operations.

After having worked the pyramidal bases of the cells of one row into their proper forms, they polish them and give them a high finish, while others are engaged in laying out the next series.

79. In fig. 33, is represented one of the faces of such a partition

Fig. 33.   Fig. 34.

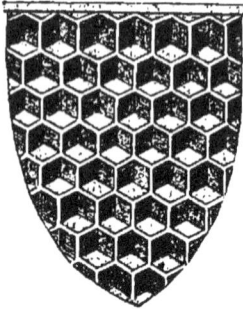 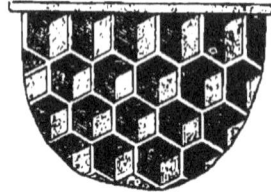

as is here described, after it has been formed into a continuous system of pyramidal bases. These are intended to represent the bases of the cells of the workers. A similar piece showing the bases of the cells of the drones is represented in fig. 34.

80. The cells themselves, consisting, as already explained, of

hexagonal prismatic tubes, are the next objects of the industry and skill of the nurses. These are cemented on the borders of the pyramidal cavities shown in figs. 26 and 27.

81. The surfaces represented in figs. 33 and 34 having a contour very unequal, the edges of the pyramidal cavities being inclined to each other, so as to form angles alternately salient and re-entrant, the first work of the bees is to form those parts of the prismatic sides of the cells which are necessary to fill up the re-entrant angles of the contours of the pyramidal bases. When this has been accomplished, the contours of all the hexagonal divisions extended over the surface of the partition, represented in figs. 33 and 34, are brought to a common level, and from that point the labour of the little artificers becomes more simple, consisting of the construction of the oblong rectangles which form the remainder of the six sides of each cell.

82. It must nevertheless be remarked, that the first row of cells, being necessarily attached to the roof of the hive, and not at its upper edge connected like the other rows with other similar cells, has an exceptional form, these being not hexagonal, but pentagonal; two of the sides of the ordinary cells being replaced by the roof of the hive, as shown in figs. 33 and 34. A corresponding exceptional form is of course also given to the bases of the first row of cells.

The combs constructed in this manner are ranged in vertical planes parallel one to the other in the hive, as shown in perspective in fig. 35, in vertical section in fig. 36, and in horizontal

Fig. 35.   Fig. 36.

section in fig. 37. They are not always ranged strictly in single parallel lines; but are sometimes bent at an angle, as shown in fig. 37.

An end view of a comb, showing the mouths of the cells foreshadowed by perspective, is represented in fig. 38.

83. The flat sides of a comb are not strictly parallel, but

39

generally slightly inclined one to the other, so that the thickness gradually diminishes from top to bottom, as shown in the vertical section, fig. 36. This gradation of thickness is continued to a

Fig. 37.   Fig. 38.

certain point, while the width of the comb is continually augmented; but so soon as the workers obtain sufficient space to lengthen it, it begins to lose this form, and the surfaces become sensibly parallel.

84. A certain class of naturalists, who have directed their attention to the history of this insect, appear to have taken a pleasure in forming hypotheses, by which it would be reduced to a mere machine. Thus, according to them, the formation of the various parts of the comb would result from a mere mechanical necessity, the organs of the insect being supposed to be so formed that the different parts of the cells would receive their forms by a mechanical process, as in certain operations in the arts the most exact geometrical forms are imparted to materials by punches and dies expressly made for the purpose.

Between such expedients and the organs of this admirable insect, there is, however, not the remotest analogy.

The mechanical instruments with which they work are the feet, the mandibles, and the tongue, the operations of which are guided by the antennæ, which are feelers of exquisite sensibility. They do not remove in their operations a single particle of wax, until the surface to be sculptured has been carefully explored by the antennæ. These organs are so flexible and so easily applied to all parts, however delicate, of their workmanship, that they are capable of performing the offices of square and compass, measuring the minutest parts with the utmost precision, so as to guide the work in the dark, and produce with unerring precision that wondrous structure called the comb.

85. It is impossible to behold a dissected comb without perceiving the geometrical necessity which connects one part with

another. In the formation of such a structure, chance can have no share. The original mass of wax is augmented by the labour of the wax-makers in the exact quantity which is necessary; and these wax-makers, who thus are constantly on the watch to observe the progress of the comb, so as to keep the artificer-bees constantly supplied with the necessary quantity of raw material, are themselves utterly destitute of the art and science necessary to construct the cells.

86. The bees never commence the construction of two contiguous and parallel combs together, for the obvious reason, as it should seem, that to make one parallel to and at a given distance from another, the actual formation of one must be first accomplished to a certain point. They therefore begin by the middle comb; and when that has been constructed to a certain depth, measured from the top of the hive, two other combs, parallel to it and at regulated distances from it at either side, are commenced; and when these again are completed to a certain depth, two others outside these are commenced, and so on. This order of proceeding is attended with a further advantage by preventing the workers on one comb from being inconveniently crowded or obtruded upon by those of the adjacent combs.

87. The labour of the bees is conducted in common, but not always simultaneously. Every partial operation is commenced by one individual bee, who is succeeded in her labours by others, each appearing to act individually in a direction depending on the condition in which she finds the work when it falls into her hands. The whole band of wax-makers, for example, is in complete inaction until one of them goes forth to lay the foundation of a comb. Immediately the labours of this one are succeeded and seconded by the others, and, when their part is done, an individual nurse-bee goes to lay out the plan of the first cell, and is in like manner succeeded continuously by others.

88. "The diameter of the cells intended for the larvæ of the workers is alway $2\frac{2}{5}$ lines, and that of those meant for the larvæ of the males or drones $3\frac{1}{2}$ lines. The male-cells are generally in the middle of the combs, or in their sides; rarely in their upper part. They are never insulated, but form a corresponding group on both sides the comb. When the bees form male-cells below those of neuters, they construct many rows of *intermediate* ones, the diameter of which augments progressively till it attains that of a male-cell; and they observe the same method when they revert from the male-cells to those of workers. It appears to be the disposition of the *queen* which decides the kind of cells that are to be made; while she lays the eggs of workers, no male-cells are constructed; but when she is about to

lay the eggs of males, the workers appear to know it, and act accordingly. When there is a very large harvest of honey, the bees increase the diameter and even the length of their cells. At this time many irregular combs may be seen with cells of twelve, fifteen, and even eighteen lines in length. Sometimes, also, they have occasion to shorten the cells. When they wish to lengthen an old comb, the sides of which have acquired their full dimensions, they gradually diminish the thickness of its edges, gnawing down the sides of the cells till it assumes the lenticular form ; they then engraft a mass of wax round it, and so proceed with new cells." *

89. The number of cells contained in the combs of a well-stocked hive is considerable. In a hive twenty inches high and fourteen inches diameter, they often amount to forty or fifty thousand. A piece of comb, measuring fourteen inches long and seven inches wide, containing about 4000 cells, is frequently constructed in twenty-four hours.

90. Nothing can be more admirable than the tender solicitude and foresight shown by the bee towards its offspring. Although these insects provide a great number of cells, as storehouses, for the honey intended for the use of the community, yet the object which more exclusively engrosses them is the care of their young, to the provision and rearing of which they sacrifice all personal and selfish considerations. In a new swarm, accordingly, the first care of these insects is to construct cradles for their young, and the next, to provide an ample store of a peculiar sort of *pap*, called *bee-bread*, for their food.

This bee-bread consists of the pollen of flowers, which the workers at this time are incessantly employed in gathering, flying from flower to flower, brushing from the stamens their yellow treasure, which they collect in the little baskets with which their hind-legs are so admirably provided. They then hasten back to the hive, where, having deposited the store thus collected, they return to seek a new load.

Another troop of labourers are in constant attendance in the hive to receive the stock of bee-bread thus collected, which they carefully store up until such time as the queen has laid her eggs. These eggs she places in an upright position in the bottoms of the cells, where they are severally hatched.

91. The bee-bread is converted into a sort of pap, or whitish jelly, by being swallowed by the bee, in the stomach of which it is probably mixed with honey and then regurgitated.

The moment the young brood issue from the eggs in the state of larvæ, they are diligently fed with this jelly by the class of bees

* Kirby, i. 419.

called nurses, who attend them with all the solicitude implied by their title, renewing the pap several times a day, as fast as it is consumed.

The curious observer will see, from time to time, different nurses introduce their heads into the cells containing the young. If they see that the stock of pap is not exhausted, they immediately withdraw and pass on to other cells; but if they find, on the contrary, the provision consumed, they never fail to deposit a fresh supply. These nurses go their rounds all day long in rapid succession thus surveying the cradles, and never stopping except where they find the supply of food nearly exhausted.

92. That the duty of these tender nurses is one which requires the exertion of some skill will be understood, when it is stated that the quality of food suitable to the young varies with their age. When they first emerge from the egg the jelly must be thin and insipid, and, according as they approach to maturity, it requires to be more strongly impregnated with the saccharine and acid principles.

Not only does the food of the larva thus require to be varied according to its age, but the food to be supplied to different larvæ is altogether different. The jelly destined for the larvæ which are to become queens, is totally different from that prepared for those of drones and workers, being easily distinguished by its sharp and pungent flavour; and it is probable, also, that the jelly appropriated to the drones differs from that upon which the workers are reared.

These insects, moreover, exhibit as much economy as skill; the quantity of food provided being as accurately proportioned to the wants of the young as its quality is to their varying functions. So accurately is the supply proportioned to the wants of the larvæ, that, when they have attained their full growth and are about to undergo their final metamorphosis into nymphs, not an atom of bee-bread is left unconsumed.

93. At the epoch of this metamorphosis, when the nymph needs seclusion to spin its cocoon, and has no further occasion for food, these tender nurses, with admirable foresight, terminate their cares by sealing up each cell, enclosing the nymph with a woven lid.

In all the maternal cares described above, neither the drones nor the queen participate. These duties fall exclusively upon the workers, and are divided between them, as has been explained, the task of collecting the bee-bread being appropriated to one set, and that of feeding and tending the young to another. This duty has no cessation; as the queen lays her eggs successively and constantly, the young arrive successively at the epoch of their first metamorphosis; and, consequently, so soon as some are sealed up and abandoned by the nurses to spin their cocoons, others issue from the egg and demand the same maternal care ; so that these nurses spend their whole existence in the discharge of the offices here described.

94. Although the organisation of other species of the bee does not approach to the perfection of the hive-bee here described, it is nevertheless worthy of attention and study.

The humble-bees, which so far as respects their social policy, compared with the hive-bee, may be regarded as rude and un-civilised rustics, exhibit nevertheless marks of affection for their young quite as strong as their more polished neighbours.

Unlike the queen of the hive, the females take a considerable share in the education of the young. When one of these provident mothers has constructed with great labour and much skill a com-modious woven cell, she furnishes it with a store of pollen moist-ened with honey, and, having deposited six or seven eggs in it, carefully closes the opening and all the interstices with wax ; but her maternal cares do not end here. By a strange instinct, pro-bably necessary to restrain an undue increase of the population, the workers, while she is laying her eggs, endeavour to seize them, and, if they succeed, greedily devour them. Her utmost vigilance and activity are scarcely sufficient to save them ; and it is only after she has again and again repelled the murderous intruders, and pursued them to the furthest verge of the nest, that she succeeds in accomplishing her object ; and even when she has sealed up the cell containing them, she is obliged to continue to guard it for six or eight hours ; since otherwise the gluttonous workers would break it open and devour the eggs. The mother is conscious, however, by a heaven-inspired knowledge, of the time when the eggs will cease to excite the appetites of the depredators.

After this the cells remain unmolested until the larva issues from the eggs. The maternal cares having there ceased, the workers, before so eager to devour the eggs, now assume the character of nurses. They know the precise hour when the larvæ will have consumed the stock of food, provided for them by maternal care, and from that time to the period of their maturity these nurses continually feed them with honey or pollen, introduced in their proboscis through a small hole in the cover of the cell opened for the purpose, and then carefully closed.

95. These nursing-workers also perform another duty of a most curious and interesting description. As the larva increases in size, the cell, which has been appropriated to it, becomes too small for its body, and in its exertions to obtain room it splits the thin woven walls which confine it. The workers, who are constantly on the watch for this, lose no time in repairing the breach, which

they patch up with wax as often as the fracture takes place, so that in this way the cell increases in size until the larva arrives at maturity.

96. As in the case of the hive-bee already described, the larva after the first metamorphosis, is shut up in the enlarged cell to spin its cocoon. When this labour has been completed, and that the perfect insect is about to issue, the workers still discharging the duty of tender foster-parents, set about to assist the little prisoner in cutting open the cocoon, from which it emerges in its perfect state.

97. While in the pupa state, however, another tender and considerate measure of the workers must not be passed without notice. It is essential to the well-being of the pupa that while concealed in the cocoon it should be maintained at a genial temperature. To secure this object, the workers collect upon the cocoons in cold weather and at night, so that by brooding over them they may impart the necessary warmth.

98. The following curious anecdote connected with this subject is related by Huber.

"He put under a bell-glass about a dozen humble-bees, without any store of wax, along with a comb of about ten silken cocoons, so unequal in height that it was impossible the mass should stand firmly. Its unsteadiness disquieted the humble-bees extremely. Their affection for their young led them to mount upon the cocoons for the sake of imparting warmth to the enclosed little ones, but in attempting this the comb tottered so violently that the scheme was almost impracticable. To remedy this inconvenience, and to make the comb steady, they had recourse to a most ingenious expedient. Two or three bees got upon the comb, stretched themselves over its edge, and with their heads downwards fixed their fore-feet on the table upon which it stood, whilst with their hind-feet they kept it from falling. In this constrained and painful posture, fresh bees relieving their comrades when weary, did these affectionate little insects support the comb for nearly three days. At the end of this period they had prepared a sufficiency of wax, with which they built pillars that kept it in a firm position: but by some accident afterwards, these got displaced, when they had again recourse to their former manœuvre for supplying their place; and this operation they perseveringly continued, until M. Huber, pitying their hard task, relieved them by fixing the object of their attention firmly on the table." *

It is impossible not to be struck with the reflection, that this most singular fact is inexplicable on the supposition, that insects are impelled to their operations by a blind instinct alone. How

* Linnæan Trans., vi. 247, et seq.

could mere machines have thus provided for a case which in a state of nature has probably never occurred to ten nests of humble-bees since the creation ? If in this instance these little animals were not guided by a process of reasoning, what is the distinction between reason and instinct ? How could the most profound architect have better adapted the means to the end—how more dexterously *shored* up a tottering edifice, until his beams and his props were in readiness ? *

99. The following remarkable example of the care bestowed by the nurses in keeping the pupa warm, more especially during the day which immediately precedes its exit from the cocoon as a perfect insect—an epoch, when as it would seem it is more especially necessary that it should be maintained at an elevated temperature,—was supplied by Mr. Newport. That naturalist observed that in the process of incubation, the humble-bee at that particular stage increased considerably the force of its respiration. To render the purpose of this intelligible to the reader not accustomed to physiological enquiries, it may be necessary to state that in the act of respiration the oxygen, which is one of the constituents of the atmosphere, enters into combination with the carbon and hydrogen, which compose part of the body of the animal. Now this combination being identical with that which produces heat in a common coal fire or the flame of a lamp, the same effect is produced in the animal economy from the same cause ; and hence it arises that the development of heat in the body is always so much the greater, in proportion to the increased activity of respiration.

100. To return to the hive-bee, it was observed by Mr. Newport that in the early stage of the incubation of the pupa, the rate of respiration of the insect is very gradual, but becomes more and more frequent as the epoch approaches at which it issues from the cocoon; the number of respirations per minute then amounting to 120 or 130.

Mr. Newport states that he has seen a bee upon the combs continue perseveringly to respire at that rate for eight or ten hours, until its temperature was greatly increased and its body bathed in perspiration. When exhausted in this way it would retire from its maternal duty and give place to another foster-mother, who would proceed in the same way to impart warmth to the pupa.

In one case Mr. Newport found that while the thermometer in the external air stood at 70·2, it rose on the lips of these cells which were not brooded upon at the moment, to 80·2, but when placed in contact with the bodies of the brooding bees, it rose

* Kirby, Int., i. 320.

to 92·5. It appears therefore that by the voluntary increase of their respiration they were enabled to impart to the nymph enclosed in the cocoon 12·3 additional degrees of heat. *

101. In every well-filled hive the combs are ranged in parallel planes, as shown in figs. 36, 37 ; and that no space may be lost, while at the same time sufficient room is left for the movements of the workers, the open spaces between the parallel combs leave a width just sufficient to allow two bees easily to pass each other. These open spaces are the streets of the apiarian city, the highways along which the building materials are carried while the combs are in process of construction, through which the supply of provisions is carried to the stores, and food to the young, who are being reared in the cells.

But since the nurses must tend the cells of all the combs, and therefore pass successively and frequently from street to street, they would be compelled to descend to the lower edge of the comb to arrive at an adjacent street, unless cross alleys were provided at convenient points to abridge such journeys. The prudent architects foresee this in laying out their city, and make such passages, alleys, or arcades, by which the bees can pass from any street to the adjacent parallel street, without going the long way round.

102. On the return of spring, when the genial temperature of the weather begins to produce its wonted effects on vegetation, and when the vernal plants which the bees love begin to put forth their foliage and flower, the busy population of the hive recommence their labours ; and the queen, who has passed the winter in repose, attended by her devoted subjects, and feeding on the stores laid up by them during the previous season, commences laying her great brood of eggs. At this epoch she is much larger than at the cessation of her laying in the autumn. Before she deposits an egg, she examines carefully the cell destined for it, putting her head and shoulders into it, and remaining there for some time, as if to assure herself that the cradle of her offspring has been put in proper order. Having satisfied herself of this, she withdraws her head, and introducing the posterior extremity of her abdomen deposits a single egg upon the pyramidal base of the cell, which adheres there in the manner already described.

She then passes to another empty cell, where, after the same precautions, she deposits another egg, and so continues, sometimes committing to the cells two hundred eggs and upwards in the day.

103. In this operation, so essential to the maintenance of the population, she is assiduously followed and most respectfully

* Philosophical Trans., 1837, p. 296.

surrounded by a certain train of her subjects, appointed apparently to attend her, and form the ladies-in-waiting on the occasion. They range themselves in a circle around her (fig. 39). From time to time

Fig. 39.—The queen depositing her eggs in the cells, surrounded by her suite.

the individuals of her suite approach her and present her with honey. They enter the cells where the eggs have been deposited, and carefully clean them, and prepare them for the reception of the pap which is to feed the young when it issues from the egg.

104. In some exceptional cases, where her majesty is rendered over prolific by any accidental cause, the eggs will drop from her faster than she can pass from cell to cell, and in such cases two or more eggs will be deposited in the same cell. Since the cells are constructed only of sufficient magnitude for the due accommodation of a single bee, the royal attendants in such cases always take away the supernumerary eggs, which they devour, leaving no more than one in each cell (fig. 40).

The eggs are oval and oblong, about the twelfth of an inch in length, of a bluish white colour, and a little bent. They are hatched by the natural warmth of the hive (from 76° to 96° Fahr.), in from three to six days, the interval depending on the temperature of the weather.

Fig. 58.—VILLAGE HIVES.

THE BEE.

ITS CHARACTER AND MANNERS.

—————

CHAPTER IV.

105. THE larva which issues from the egg is a white grub, destitute of legs, having its body divided transversely by a series of parallel circular grooves into annular segments. When it has

grown so as to touch the opposite angle of the cell, it coils itself up in the form of a circular arc, or as Swammerdam describes it, like a dog going to sleep. It floats there in a whitish transparent fluid, provided for it by the nurses, on which it probably feeds during this early stage of its life. Its dimensions are gradually enlarged until its extremities touch one another, so as to form a complete ring, fig. 41, in the base of the cell. In this state the grub is fed with the pap or bee bread already mentioned. The slightest movement on the part of the nursing bees is sufficient to attract its attention, and it eagerly opens its little jaws to receive the offered nourishment, the supply of which, presented by the nurse, is liberal without being profuse.

Fig. 40.

Fig. 41.

Fig. 42.

Fig. 43.

The growth of the larva is completed in from four to six days, according to the temperature of the weather. In cool weather the development takes two days more than in warm weather.

When it has attained its full growth, it occupies the whole breadth and a great part of the length of the cell. The nurses at this time knowing that the moment has arrived at which the first metamorphosis, in which the grub is changed into a nymph, takes place, discontinue the supply of food, and close up the mouth of the cell by a light brown waxen cover, which is convex externally. This convexity of the cover is greater in the drone cells than in those of the workers. The covers of the honey cells are, on the contrary, made paler in colour, and quite flat or even a little concave externally.

Fig. 44.

When the larva has been thus enclosed, it immediately commences, like the silk-worm, to spin a cocoon. In this labour it is incessantly employed, lining the sides of its cell and encasing its own body in a white silken robe. The threads which form this mantle issue from the middle of the under lip of the nymph, as the insect in this intermediate state between that of the grub and the perfect bee is called. This thread consists of two filaments, which, issuing from two adjoining orifices in the spinner, are then gummed together.

106. The nymph of a worker spins its robe in thirty-six hours, and after passing three days in this preparatory state, it undergoes so great a change as to lose every vestige of its previous form. It

is clothed with a harder coating, with dark brown scales, fringed with light hairs. Six annular segments are distinguished on its abdomen, which are inserted one into another like the joints of a telescope tube, and give the insect the power of elongating and contracting itself within certain limits. The breast is also invested with a sort of brush of grey feathery hairs, which as age advances assume a reddish hue. In about twelve days all the parts of the body of the perfect insect are developed, and can be seen through the semi-transparent robe in which it is clothed.

About the twenty-first day, counting from that on which the egg was laid, the second metamorphosis is complete, and the perfect insect, gnawing through the cover of its cell, issues into life, leaving behind it the silken robe which it wore in the intermediate state of nymph. This is closely attached to the inner surface of the cell in which it was woven, and forms a permanent lining of it. By this cause the breeding cells become smaller and smaller, as the eggs are successively hatched in them, until at length their capacity becomes too limited for the full development of the nymphs. They are then turned into store rooms for honey.

Fig. 45.

Pupa of a worker.

107. In fig. 46 is represented a piece of comb, consisting exclusively of workers' cells, in different states. Several, c, c, c, &c., are closed, the nymph not having yet undergone its final metamorphosis. A bee having arrived at the perfect state and gnawed open the

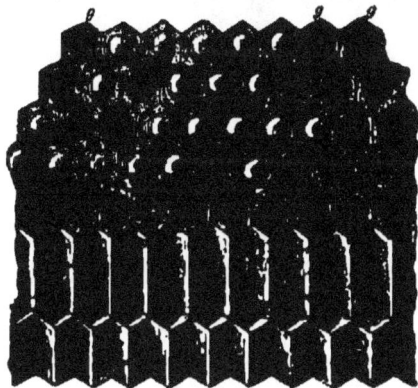

Fig. 46.

cover of its cell, is shown at m. The cells, h, h, have their openings on the opposite side of the comb, and g, g, g, are cells from which the perfect insects have already issued.

108. When a young bee, after its final metamorphosis, has issued from the cell, the nurses crowd round it, carefully brushing it, giving it nourishment and showing it the way through the hive. Others meanwhile are occupied in cleaning the cell from which it has issued and putting it in order to receive another egg if it be still large enough, and if not, to receive a store of honey.

The young bee is not sufficiently strong to fly on the first day. It is only on the morrow, after being well fed and brushed down by the nurses, and having taken a walk from time to time through the combs, that it ventures on the wing.

109. The drone passes three days in the egg, and continues to receive the care of the nurses as a grub until the tenth day, when it passes into the state of nymph, and is sealed up in its cell by the

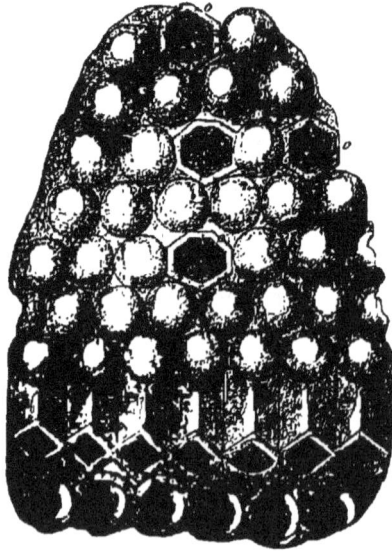

Fig. 47.

nurses with a very convex cover. As already stated, the drone grub being larger than that of the worker, the cell assigned to it is proportionately more capacious, and the cover by which as a nymph it is shut up is much more convex externally. A piece of comb consisting of drone cells is shown in fig. 47.

Some cells, *a, o, o,* being those from which the perfect insect has issued, are open and empty.

Near the borders of the comb, where local circumstances render it necessary to modify the principles of its architecture so as to accommodate the cells to their position in the hive, may be observed several, k, k, of unusual and irregular forms. While some such cells have six unequal sides, others have only four or five. It seems also that in the case of certain cells intended only for the reception of honey, the bee is not at all as scrupulous in the observance of architectural regularity as in the case of brood cells.

110. The drone nymph undergoes its final metamorphosis and becomes a perfect insect, from the twenty-fifth to the twenty-seventh day from that on which the egg is laid, according to the temperature of the hive. It is therefore six or seven days later in arriving at maturity than the worker.

111. The changes to which the young of the royal family are subject before arriving at maturity, are different from those above stated. It has been already explained that the royal cells are vertical instead of being horizontal, are egg-shaped instead of being hexagonal, and in fine are much more capacious than those

Fig. 48.

of the drones or workers. One of these cells is shown at r s in fig. 48, a part, u u, being removed to show the royal nymph within it. It will be observed that a much larger space is given to the royal nymph than is allowed either to that of the worker or the drone, the bodies of which nearly fill their respective cells. The royal nymph is always placed, as shown in the figure, with her head downwards.

The progressive formation of a royal cell is shown in fig. 49. It is unfinished, as at a, when the egg is deposited; and is gradually enlarged, c, as the grub increases in size; and is sealed up, b, when it is transformed into a nymph.

53

The grub issues from the egg on the third day, becomes a nymph from the eighth to the eleventh day, and undergoes its

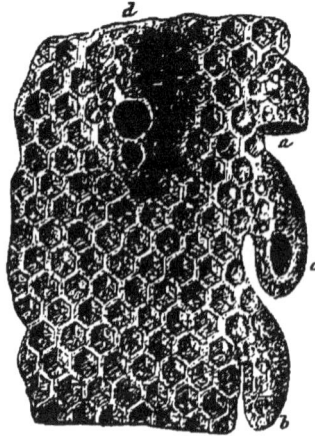

Fig. 49.

final metamorphosis, becoming a perfect insect on the seventeenth day. It is, however, sometimes detained a prisoner in the cell for seven or eight days longer.

112. Naturalists are not agreed as to some of the circumstances attending the treatment of the young, which we have here given on the authority of Feburier and other French entomologists. Mr. Dunbar, in reference to the circumstances attending the first issuing of the perfect insect from the cell, says that in hundreds of instances their situation has excited his compassion, when after long struggling to escape from its cradle, it has at last succeeded so far as to extrude its head, and when labouring with the most eager impatience, and on the very point of extricating its shoulders also, which would have at once secured its exit, a dozen or two of workers, in following their avocations, have trampled without ceremony over the struggling creature, which was then forced for the safety of its head, quickly to pop down again into the cell and wait until the unfeeling crowd had passed, before it could renew its efforts. Again and again will the same impatient efforts be repeated by the same individual, and with the same mortifying interruptions, before it succeeds in obtaining its freedom. Not the slightest attention or sympathy on the part of the workers in these cases was ever observed by Mr. Dunbar, nor did he ever witness the parental cares and sage instruction given to the young which are described by the French entomologists.

Positive, however, is more entitled to consideration than negative testimony, and it cannot be doubted that Feburier and others witnessed those cares, guidance, and education which they have so well described. Besides, Dr. Bevan admits that he has seen assistance rendered to the infant drones. So soon as the young insect has been cleaned of its exuviæ and regaled with honey by the nurses, the latter clean out the cell exactly as we have already described.

113. A piece of comb is shown in fig. 50, the upper part A, of which contains honey-cells closed with flat sides of wax. The cells, c c, &c., contain pollen, and c′ c′, &c., propolis. The cells

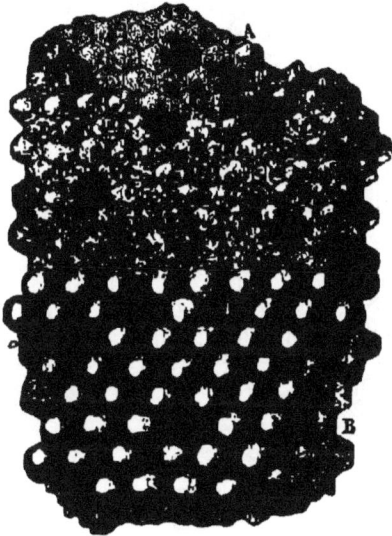

Fig. 50.

of the upper part are those which originally belonged to workers, and those of the lower part, with convex covers, are occupied by the drone nymphs.

114. The various flowers and herbs which supply the materials for honey, wax, and propolis taken collectively, are called the pasturage of the bees, and it is observed that when this pasturage is very abundant, the bees, eager to profit by the rich harvest, depart from their habit of conveying their booty first to the uppermost cells of the comb, so as to fill them gradually downwards. On the contrary, upon arriving with their load, and eager to return for a fresh supply, they unload themselves in the nearest empty cells they can find. The wax-makers meanwhile charge

themselves with the labour of taking the provisions thus deposited from the lower to the upper parts of the combs.

115. In fig. 51, is shown a piece of comb in process of construction. It has, as usual, an oval form. The wax, of which it is formed, is white, but as it advances in age it takes successively a

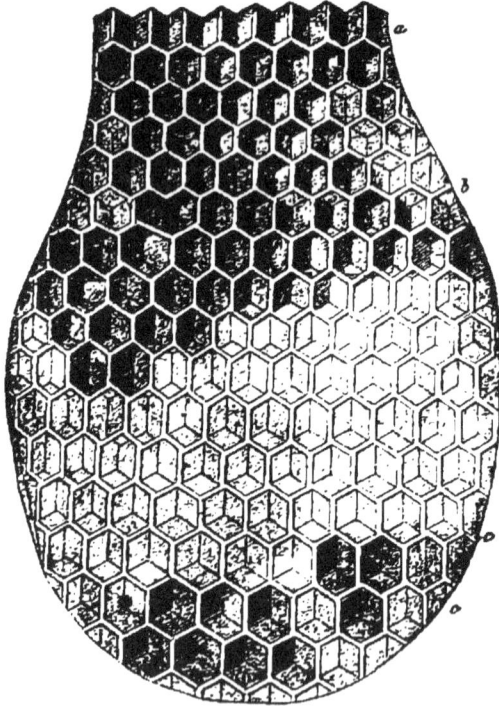

Fig. 51.

darker and darker colour, being first yellow, then reddish, and sometimes even becomes blackish. The sides of the cells are gradually thickened, by the constant adhesion and accumulation of the cocoons, of which the nymphs successively bred in them are divested. The top and sides of the comb are every where strongly cemented, by a mixture of propolis and wax, to the roof and sides of the hive. These structures are almost never known to fall except by some accidental cause external to the hive, such as a blow or the too intense heat of the sun dissolving the cement.

116. The character and manners of the bee have an intimate relation with its social organisation. We have seen that in the

56

building of their city this organisation is never for a moment lost sight of. The chambers vary in number, magnitude, form, and position. Those designed for the members of the royal family are few and exceptional, those for the drones much more numerous, but about one hundred times less numerous than those of the workers. The magnitudes are in like manner strictly regulated, in relation to the volume of the body of the occupant, except the royal chambers to which a magnitude is given much greater in proportion than that of the bodies of the royal tenants. The object to be attained by this increased capacity, as well as by the vertical position specially given to the royal cells, has not been ascertained.

117. How little relation there exists between mere bodily magnitude, and the faculties which govern acts so remarkable as those of the insects now before us, will be understood when it is stated that, according to the experiments of Reaumur, the average weight of the bee is such that 336 go to an ounce, and 5376 to a pound ; and John Hinton found that 2160 workers would not more than fill a common pint.

118. Having thus explained in a general way the persons composing the society, and the structure and architecture of their dwellings, we shall proceed to notice some of the more remarkable traits of their character and manners.

It has been already explained that the community of the hive bees is strictly a female monarchy. The jealous Semiramis of the hive, as Kirby observes, will have no rival near her throne. It may, therefore, be asked to what purpose are the sixteen or twenty princesses reared, for whom royal chambers are provided, and who are treated in all respects by the nurses as aspirants to the throne? This will be comprehended, however, when it is remembered that the hive, soon after the commencement of the season, becomes so enormously over-peopled, that emigration becomes indispensable, and that with each emigrant swarm a queen is necessary. Either therefore the queen regnant must go forth, abdicating the throne, in which case it is ascended by the eldest of the princesses, or the latter is raised to the sovereignty of the emigrating colony. Now, since a rapid succession of swarms issue from the hive, especially in the early part of the season, sometimes as many as four in eighteen days, and since one queen is required for each, a proportionately numerous royal family is required to fill so many independent thrones.

119. When the growth of several princesses and their arrival at maturity occurs, before the increase of the population renders emigration necessary, so as to create thrones for them, the most violent jealousy is excited in the breast of the queen regnant, who is either mother or sister to these several queens presumptive,

57

and her royal breast is fired with agitation, nor does she rest until she has engaged in mortal conflict with her rivals, and either puts them to death or suffers death at their hands.

120. When a hive, having lost its queen by emigration or otherwise, is provided with several royal cells, which generally happens, the first princess which issues from these in the perfect state immediately ascends the throne in right of primogeniture. Although her rivals are not yet in a condition to dispute the title, they, nevertheless, excite her jealousy in the highest degree. Scarcely ten minutes elapse from the moment she has attained the perfect state, and issued from the royal cell, when she goes in quest of the other royal cells, assails with fury the first she encounters, and having gnawed a large hole in it she introduces the posterior extremity of her abdomen, and kills her rival with her sting.

121. A crowd of workers, who are passive spectators of this, approach the cell, and enlarging the breach, drag out the corpse of the murdered princess, who, in such cases, has already assumed the perfect state. If the queen attack in like manner a cell of which the occupant is still in the state of nymph, she does not waste her strength in slaying it, well knowing that its premature exposure will do the work of death. The workers, in this case also enlarging the breach made by the queen, pull out the nymph, who immediately perishes.

122. Huber, who witnessed, and has described all these curious proceedings, being desirous to ascertain what would happen if two rival queens, both in the perfect state, found themselves together in the same hive, produced artificially that contingency on the 15th May, 1790. He managed to provide in the same hive royal cells, in an equal stage of forwardness, so that virgin queens issued from two of them almost at the same moment.

When they appeared in presence of each other they fell upon each other with all the appearance of insatiable fury, and so engaged one with the other, that each held in her mandibles the antennæ of the other. They were engaged breast to breast, and abdomen to abdomen, so that if each had put forth her sting, mutual death would have been the consequence. But as if nature had forbidden this mutual destruction, the combatants disengaged themselves from each other's grasp, and fled one from the other with the greatest precipitation.

Huber says that this was not a mere incident which might have occurred in a single case, but would not occur in others, for he repeated the same experiment frequently, and it was always followed by the same result. It seemed, therefore, as though it were a case foreseen by nature, and that one only of the combatants should fall in such combats.

123. Nature has ordained that in each hive there shall be one, and but one queen, and when by any concurrence of circumstances a second appears, one or the other is doomed to destruction. But it is not permitted to the common class of the people to do execution on a royal personage, since in that case it might not be possible to secure unanimity as to the particular queen who is to be preserved, so that different assemblages of the people might at the same time assail different queens, and so leave the hive without a sovereign. It was, therefore, necessary, as Huber argues, that the extermination of the superfluous queens should be left to the queens themselves, and that they should in their combats be filled with an instinctive horror of mutual destruction.

Some minutes after the two queens above mentioned had separated and retired from each other, and when their fears had time to subside, they again prepared to approach each other. They engaged once more in the same position, involving the danger of mutual destruction, and as before, once again separated and mutually fled each other.

124. During all this time the greatest agitation prevailed among the population who assisted at the scene, more especially when the two combatants separated. On two different occasions the workers interfered to prevent them from flying from one another. They arrested them in their flight, seizing them by the legs and detaining them prisoners for more than a minute. In fine, in a last attack, one of the queens, more active and furious than the other, taking her rival unawares, laid hold of her with her mandibles at the insertion of the wing, and then mounting on her back, and bringing the posterior extremity of her abdomen to the junction of one of the abdominal segments of her adversary, stabbed her mortally with her sting. She then let go the wing which she had previously held and withdrew her sting.

The vanquished queen fell, dragged her body slowly along for a certain distance, and soon after expired.

125. Having thus ascertained the conduct of virgin queens under the circumstances here described, Huber made arrangements for observing the conduct of queens who were in a condition to produce eggs. For this purpose he placed a piece of comb on which three royal cells had been constructed in a hive with a laying queen. The moment they caught her eye she fell upon them, opened them at their bases, and surrendered them to the attendant workers, who lost no time in dragging out the royal nymphs, greedily devouring the store of food which remained in the cells, and sucking whatever was in the carcases. Having accomplished this they proceeded to demolish the cells.

It was now resolved to ascertain what would be the behaviour of

59

a queen-mother regnant in case a stranger queen pregnant were introduced into the hive. A mark having been previously made upon the back of such a queen, so that she might be afterwards identified, she was placed in the hive. Immediately on her appearance the workers collected in a crowd around her, and formed as usual a circle of which she was the centre, the heads of all the remaining crowd being directed towards her. This very soon became so dense that she became an absolute prisoner within it.

While this was going on, a similar ring was formed by another group of workers round the queen regnant, so that she was likewise for the moment a prisoner.

The two queens being thus in view of each other, if either evinced a disposition o approach and attack the other, the two rings were immediately opened, so as to give a free passage to the combatants; but the moment they showed a disposition to fly from each other, the rings were again closed, so as to retain them in the spot they occupied.

At length the queen regnant resolved on the conflict, and the surrounding crowd, seeming to be conscious of her decision, immediately cleared a passage for her to the place where the stranger stood perched on the comb. She threw herself with fury on the latter, seized her by the root of the wing, and fixed

Fig. 52.

her against the comb so as to deprive her of all power of movement or resistance, and then bending her abdomen inflicted a mortal stab with her sting, and put an end to the intruder.

126. A fruitful queen full of eggs was next placed upon one of the combs of a hive over which a virgin queen already reigned. She immediately began to drop her eggs, but not in the cells; nor did the workers, by a circle of whom she was closely surrounded, take charge of them; but, since no trace of them could be discovered, it is probable that they were devoured.

The group, by which this intruding queen was surrounded, having opened a way for her, she moved towards the edge of the comb, where she found herself close to the place occupied by the legitimate virgin queen. The moment they perceived each other, they rushed together with ungovernable fury. The virgin, mounting on the back of the intruder, stabbed her several times in the abdomen, but failed to penetrate the scaly covering of the segments. The combatants then, exhausted for the moment, disengaged themselves and retired. After an interval of some minutes they returned to the charge, and this time the intruder succeeded in mounting on the back of the virgin and giving her several stabs with her sting, which, however, failed to penetrate the flesh. The virgin queen, succeeding in disengaging herself, again retired. Another round succeeded, with the like results, the virgin still coming undermost, and, after disengaging herself, again retiring. The combat appeared for some time doubtful, the rival queens being so nearly equal in strength and power, when at last, by a lucky chance, the virgin sovereign inflicted a mortal wound upon the intruder, who fell dead on the spot.

In this case, the sting of the virgin was buried so deep in the flesh of her opponent, that she found it impossible to withdraw it, and any attempt to do so by direct force would have been fatal to her. After many fruitless efforts she at length adopted the following ingenious expedient with complete success. Instead of exerting her force on the sting by a direct pull, she turned herself round, giving herself a rotatory motion on the extremity of her abdomen where the sting had its insertion, as a pivot. In this way she gradually *unscrewed* the sting.

127. The gates of the hive are as constantly and regularly guarded night and day as those of any fortress. The workers charged with this duty are, of course, regularly relieved. They scrupulously examine every one who desires to enter; and, as though distrustful of their eyes, they touch all visitors with their antennæ. If a queen happens to present herself among such visitors, she is instantly seized and prevented from entering. The sentinels grasp her legs or wings with their mandibles, and so surround her that she cannot move. As the report of the event spreads through the interior of the hive, large reinforcements of the guard arrive, who augment the dense ranks which hold the strange queen in durance.

In general, in such cases, the intruding queen is thus detained prisoner until she dies from want of food. It is remarked that the guard, who thus surround and detain her, never use their stings upon her. In one instance Huber attempted to extricate a queen, thus surrounded, by taking her directly out of the ring of guards. This excited the rage of the guard to such a pitch that, putting forth their stings, they rushed blindly not only on the queen but on each other. The queen, as well as several of the guard, were killed in the mêlée.

128. When the sovereign of the hive is removed or accidentally destroyed, the population seem at first to be wholly unconscious of their loss, and pursue their usual avocations as if nothing had happened. But after the lapse of some hours they begin to manifest a certain degree of uneasiness. This gradually increases,

until the entire hive becomes a scene of tumult. The wax-makers abandon their work, the nurses desert the infant brood; they run here and there in all directions through the streets and passages of the hive as if in delirium. That all this disorder and alarm is produced by the report spreading that the sovereign has disappeared, was proved to demonstration by Huber, who restored to the hive the queen he had purposely withdrawn. Her majesty was instantly recognised by those who happened to be assembled at the place of her restoration ; but what is remarkable is that the intelligence of her return was immediately spread through every part of the hive, so that the bees in its most remote streets and alleys, who had no opportunity of personally seeing her majesty, were informed of her re-appearance, as was proved by the restoration of order and tranquillity, and the resumption of their usual labours by all classes of the population.

129. If, instead of restoring to the hive the queen herself, a new queen, stranger to the population, be introduced, she will not at first be accepted. She will, on the contrary, be guarded and imprisoned by a ring of bees, in the same manner as a strange queen is treated in a hive which still retains its reigning sovereign. But if she survives sixteen or eighteen hours in this confinement, the guard around her gradually disperses itself, and the lady enters the hive and assumes without further question the state and dignity of queen, and becomes the object of the homage paid to the sovereign.

As we have already stated, the first work which the population undertakes, after being assured of the loss of its queen, is directed to obtain a successor to her. If there be not royal cells prepared, they set about their construction. While this work was in progress, and in twenty-four hours after their queen had been taken from them, Huber introduced into the hive a fruitful queen in the prime of life, being eleven months old. Not less than twelve royal cells had been already commenced and were in a forward state. The moment the strange queen was placed on one of the combs, one of the most curious scenes commenced which was probably ever witnessed in the animal world, and which has been described by Huber.

The bees who happened to be near the stranger approached her, touched her with their antennæ, passed their proposces over all parts of her body, and presented her with honey. Then they retired, giving place to others, who approached in their turn and went through the same ceremony. All the bees who proceeded thus clapped their wings in retiring and ranged themselves in a circle round her, each, as it completed the ceremony, taking a position behind those who had previously offered their respects. A

62

general agitation was soon spread on those sides of the combs corresponding with that of the scene here described. From all quarters
the bees crowded to the spot, and each group of fresh arrivals
broke their way through the circle, approached the new aspirant
to the throne, touched her with their antennæ and probosces,
offered her honey, and, in fine, took their rank outside the circle
previously formed. The bees forming this sort of court circle
clapped their wings from time to time, and fluttered apparently
with self-gratification, but without the least sign of disorder or
tumult.

At the end of fifteen or twenty minutes from the commencement of these proceedings the queen, who had hitherto remained
stationary, began to move. Far from opposing her progress or
hemming her in, as in the cases formerly described, the bees
opened the circle on the side to which she directed her steps,
followed her, and, ranging themselves on either side of her path,
lined the read in the same manner as is done by military bodies
in state processions. She soon began to lay drone eggs, for which
she sought and found the proper cells in the combs which had
been already constructed.

While these things were passing on the side of the comb where
the new queen had been placed, all remained perfectly tranquil
on the opposite side. It seemed as though the bees on that side
were profoundly ignorant of the arrival of a new queen on the
opposite side. They continued to work assiduously at the royal
cells, the construction of which had been commenced on that side
of the comb, just as if they were ignorant that they had no
longer need of them; they tended the grubs in those cells where
the eggs had been already hatched, supplying them as usual,
from time to time, with Royal Jelly. But at length the new
queen in her progress arriving at that side of the comb, she was
received by those bees with the same homage and devotion of
which she had been already the object at the other side. They approached her, coaxed her with their antennæ and probosces, offered
her honey, formed a court circle round her when she was stationary,
and a hedge at either side of her path when she moved, and proved
how entirely they acknowledged her sovereignty by discontinuing
their labour at the royal cells, which they had commenced before
her arrival, and from which they now removed the eggs and
grubs, and ate the provisions which they had collected in them.

From this moment the queen reigned supreme over the hive,
and was treated in all respects as if she had ascended the throne
in right of inheritance.

130. Most of the proceedings of these curious little societies are
explicable by what seems a general social law among them, to

suffer no individuals or class to continue to exist, save such as are necessary in one way or another to the well-being of the actual community, or the continuance of the species. This principle once admitted, we find explanations satisfactory enough of all the circumstances attending the conduct of the queen regnant towards the royal princesses, of the population generally to the several members of the royal family, and, in fine, of the workers towards the drones.

The royal family, as we have seen, are all fertile females, and their sole function is to assume the throne of the hive itself, or of the colonies called swarms, which successively issue from it, and thus placed to become the fruitful mothers of thousands, which will continue the race and form future colonies.

The drones have no other function than that of kings consort presumptive, either of the hive itself or of the colonies which successively emigrate from it. As has been explained, one only is chosen as consort by each queen. So long as the swarming season continues, a sufficient body of drones are wanted to supply the necessary troop of suitors to each emigrant princess. But when the last swarm of the season has gone forth, and the queen regnant has long since made her choice and celebrated her nuptials, the drones are no longer useful to the general population, and become the objects of a general massacre.

131. After the close of the winter, and at the commencement of the first fine days of spring, the active life of the society recommences. A well peopled hive is then always provided with a fertile queen, who has held the sovereignty since the close of the preceding season. In the months of April and May she begins to lay drone eggs in great numbers. This is called the great laying.

While she is thus engaged depositing her eggs in the larger class of hexagonal cells, previously constructed for their reception, the workers, well knowing that the deposition of royal eggs will speedily follow, occupy themselves in constructing a number of those cells of oval shape and vertical position, (fig. 49,) which have been already described.

Fig. 56.—THE CABINET BEE-HOUSE.

THE BEE.

ITS CHARACTER AND MANNERS.

—◆—

CHAPTER V.

To make this great laying of drone eggs, her majesty must be at least eleven months old. Supposing that she has been hatched the preceding season in February, she will lay during that season workers' eggs almost exclusively, producing at the most from fifty to sixty drone eggs. But after the winter, at the epoch now referred to, the hive being then filled exclusively with workers, and standing in absolute need of drones to supply suitors to the future queens, she produces drone eggs constantly and exclusively until the commencement of the swarming season, with the exception, however, of a limited number of royal eggs, which she deposits at intervals more or less distant in the royal cells just now mentioned, which the workers occupy themselves in constructing during the great laying.

The great laying usually continues for about a month, and it is about the twentieth or twenty-first day that the workers begin to lay the foundations of the royal cells. They generally build from sixteen to twenty of them, and sometimes even as many as twenty-seven. When these cells have attained the depth of two-tenths to three-tenths of an inch, the queen deposits in each of them successively a royal egg. Now since the princesses which are to issue from these eggs are destined to ascend the thrones of the emigrant colonies, which are to issue in succession from the hive, it is important that they should arrive at maturity at successive intervals, corresponding as nearly as possible with the emigration of the swarms.

The queen acts as if she were conscious of this, for she deposits the royal eggs, not like the drone or worker eggs in rapid and uninterrupted succession, but after such intervals as will insure their arrival at maturity in that slow succession, which will correspond nearly or exactly with the issue of the successive swarms.

132. It has been already explained that the nurses seal up the cells, at the time at which the grub is ready to undergo its metamorphosis into a nymph. In accordance with this, and with the successive deposition of the royal eggs, just described, the times of sealing up the series of royal cells are separated by intervals corresponding with those of the deposition of the royal eggs.

Before the commencement of the great laying, the abdomen of the queen is so enlarged that her movements are seriously impeded, and she would be altogether unable to fly. According as the laying proceeds, she becomes smaller and smaller, and when it has been completed, the royal eggs having been meanwhile deposited at regulated intervals, as above described, her majesty recovers her natural form and dimensions, and with them her full bodily activity. This change in the condition of the queen, and

66

the simultaneous deposition of fifteen hundred to two thousand drone eggs, and some sixteen or twenty royal eggs, are intimately connected with the approaching social state of the colony.

133. It was shown by Huber, and since confirmed by other observers, that it is a constant law of bee politics that the first swarm of the season shall be led by the queen-regnant, who therefore abdicates her native throne in favour of the colonial sovereignty. This swarm takes place when the grub proceeding from the first of the eggs deposited by the queen in the royal cells, as above described, has undergone its transformation into a nymph.* The necessity for this law is thus explained by Huber. Without it, the mutual conflict of the queen-regnant and the princesses, as they would be successively developed, would render the emigration of swarms impossible. For as each princess would issue perfect from the cell, she would be attacked, and forced to engage in combat with the queen, who being, by reason of her age, the stronger and more powerful, would be always victorious. Thus princess after princess would be destroyed, and none would be forthcoming to take the thrones of the successive emigrating colonies. To prevent such a catastrophe, nature has therefore wisely ordered that the queen-regnant, by leading forth the first swarm of the season, should remove all cause of danger to the succession of princesses.

134. When the emigrant swarm thus first sent forth from the parent hive has established itself, the first care of the workers is to construct combs, consisting of workers' cells. They labour assiduously at these, and in accordance with this the queen, who has already deposited in the original hive her full brood of drone eggs, soon begins in her new city to deposit a brood of worker eggs; workers being then the first and most pressing want of the colony. This laying begins as soon as the cells are ready for the deposition of the eggs, and continues for ten or twelve days. About the latter part of this interval, the bees occupy themselves in the construction of the larger class of hexagonal cells for the drone eggs. It would seem as though they knew that her majesty would at this time lay a certain number of such eggs. She accordingly commences laying these, though in far less number than in the great laying, but still sufficient to prepare her people for the succeeding deposition of royal eggs, for which they construct meanwhile a suitable number of royal cells.

It rarely happens, at least in the country where Huber made his observations, that the original queen leads forth a swarm from the new hive. The thing nevertheless occasionally occurs, and when it does, it takes place in three or four weeks after the

---

* Huber, i. 279.

original swarm, and is attended with circumstances precisely similar.

135. Let us now return to the original hive and see what took place there after the departure and abdication of the reigning queen.

As examples proving the loyalty and fidelity of the bees to their queen, Dr. Bevan quotes some remarkable and interesting cases supplied by Dr. Warder. That apiarist being desirous of ascertaining the extent of the loyal feeling among these little people, hazarded the loss of a swarm in an experiment made with that object. Having shaken on the grass all the bees from a hive which they had tenanted only the preceding day, he carefully sought for and quietly caught the queen. Then placing her with a few attendants in a box, he took her into his parlour, where the lid being removed, she and her attendants immediately flew to the window, when he clipped off one of her wings, returned her to the box and confined her there for more than an hour.

In less than a quarter of an hour the swarm ascertained the loss of their queen, and instead of clustering together in a single mass as usual, like a bunch of grapes, they spread themselves over a space of several feet, were much agitated, and uttered a plaintive sound. An hour afterwards they all took flight and settled upon the hedge where they had first alighted after leaving the parent stock, but instead of clustering together in a single bunch, as when the queen accompanied them, and as swarms usually hang, they extended themselves thirty feet along the hedge in small bunches of forty or fifty or more.

The queen was now presented to them, when they quickly gathered round her with a joyful hum, and formed one harmonious cluster. At night the Doctor hived them again, and on the next morning repeated the experiment to see whether the bees would rise. The queen being in a mutilated state, and unable to accompany them, they surrounded her for several hours apparently willing to die with her rather than abandon her in her distress. The queen was a second time removed, when they spread themselves out again, as though in search of her. Her repeated restoration to them at different parts of their circle produced one uniform result, and these poor loving and loyal creatures always marched and counter-marched every way as the queen was laid. The Doctor persevered in these experiments, till, after five days and nights of voluntary fasting, they all died of inanition except the queen, and she survived her faithful subjects only a few hours.

This remarkable attachment between queen and subjects appears to be reciprocal, the sovereign being as strongly sensible of it as

thoso over whom she rules. Though offered honey on several
occasions during her temporary separation from the swarm in
these experiments, she constantly refused it, disdaining a life
which was no life to her, deprived of the society of her faithful
people.*

136. After the departure of her majesty there seems to be a sort
of interregnum in the hive during the succession of swarms.   No
new sovereign is for the moment elevated to the throne.   A strong
guard is established at each of the royal cells, whose duty it is
to confine the princesses with the utmost rigour to their respective
cells, carefully feeding them, and only liberating them at intervals
of some days according to the successive departure of the swarms.
They are liberated in the strict order of their seniority, the
nymph proceeding from the first royal egg, or the princess royal,
being invariably the first set free.

137. When she issues forth, her first impulse, like that of all
queens, is to fall upon the cells containing her younger sisters to
destroy them. This, which in other states of the colony is permitted
by the workers, is now strenuously and effectually opposed by them.
When she approaches the neighbourhood of the royal cells, the
guard in whose charge these are placed, pinch, worry, and hunt
her until they compel her to depart, but never attempt to assail
her with their stings or seriously injure or disable her.

Now, as there are usually a great number of these royal cells in
different parts of the hive, our princess finds it a difficult matter
to obtain any corner where she can remain unmolested.   Incessantly impelled by her instinct to attack the cells of her sisters,
and as incessantly repulsed from them by the surrounding guard,
her life is rendered miserable.   She is in a constant state of
agitation, running from one group of workers to another, until at
length the agitation is shared by a certain portion of the workers
themselves.   When this occurs, a crowd of bees are seen rushing
towards the portals of the city.   They issue from it accompanied
by their young and virgin queen.   It is the second swarm of the
season, and differs from the first only in the age and condition of
its sovereign.

138. After this emigration the workers, who have remained in
possession of the hive liberate another of the princesses, the
second in seniority, whom they treat exactly in the same manner
as the former.   The same succession of repulses by the guards of
the remaining royal cells takes place, attended by like consequences,
this second princess leading forth in the same manner the third
swarm, and so on.

139. This spectacle is repeated three or four times in the season

* Bevan, p. 148.

69

in a well-peopled hive, until the population is so reduced that the number necessary to form a sufficient guard upon the royal cells ean no longer be spared from the general industry of the hive. Several princesses then escape from the cells, nearly at the same time, who fall upon each other in the manner already described, being now encouraged instead of being opposed by the workers. In fine, all but one fall in those combats, and this fortunate survivor, who is in general the eldest of the princesses remaining in the hive, ascends the throne, and is acknowledged by the whole community.

According to Huber, swarms issue from the hive only in sunshine and a calm atmosphere. After all the precursors of a swarm have appeared, a passing cloud often arrests it, and the intention of the bees seems to be abandoned. An hour later the appearance of the bright sun will reproduce all the usual movements, and the swarm will issue.

Many conjectures are made as to the means by which the workers know so well, as they undoubtedly do, the relative ages of the several princesses, so as to liberate them according to seniority. Huber conjectures that a peculiar sound, which they produce before their liberation from the cells, and which he thought varied in loudness and pitch, might be the distinguishing character of relative age.

140. A contingency arises occasionally in the bee community, which we have not yet noticed, and which is attended with consequences of a very curious and interesting nature. It was discovered by Schirach, and confirmed by numerous and long continued observations of Huber, that when by any cause a colony loses its queen, without having any royal cells or royal eggs previously provided, they are enabled by certain extraordinary processes and expedients to produce princesses, among whom they may obtain a successor to their last sovereign.

M. Schirach, Secretary of an Apiarian society, at Little Bautzen in upper Lusatia, observed that bees, when shut up with a portion of comb containing worker brood only, would soon construct royal cells, into which they would put worker eggs, the grubs from which, being nourished with royal jelly, would grow up as queens. This remarkable result is known among apiculturers as the Lusatian experiment. This experiment has since been repeated thousands of times, and always with the same results by all the most eminent naturalists who have directed their researches to this part of entomology, and indeed generally by all bee cultivators. So that of the fact itself, strange and incredible as it may seem, there is not the faintest shadow of doubt.

70

The following is the process by which this miracle of nature is performed.

Having chosen a worker grub, from one to three days old, the workers pull down two of the cells adjacent to that in which the chosen grub lies. They pull down the walls which separate these three chambers, so as to throw them into one three times more spacious than the single cell of the grub. Leaving the pyramidal bases of these three cells untouched, they construct around the grub a large cylindrical tube, which is consequently included within the remaining walls of the three demolished cells, the axis of the tube being parallel to that of the cells, and therefore horizontal.

It seems, however, that to accomplish the desired change on the nature of the grub, it is not only necessary to give it an enlarged cell, but one of which the axis is vertical instead of being horizontal. On the third day, therefore, from the commencement of their operations, they take measures to cement to the horizontal tube a vertical chamber having a conical form, making with the horizontal tube an elbow. To accomplish this they gnaw away several cells below the end of the tube, sacrificing without mercy the grubs which occupy these, as well as those which occupied the two cells adjacent to the original cell of the chosen grub.

This rectangular cell, therefore, composed of the original cylindrical, and the more recently constructed conical cell, may be considered as having some such form as here roughly sketched,

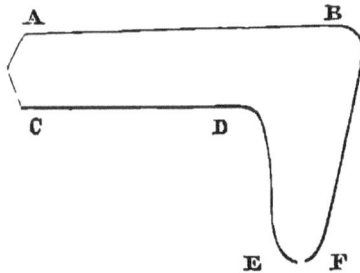

Fig. 53.

(fig. 53,) where A B C D is the horizontal cylindrical part formerly filled by three worker hexagonal cells, and B F E D, the vertical conical part, subsequently cemented to it, and built with the wax obtained from the demolition of the worker cells under A B C D.

During two days which the grub inhabits this vertical cell, B F D E, a nurse may always be observed with its head plunged

71

into it, and when one quits it another takes its place, thus relieving each other with all the regularity of military sentinels. These bees keep constantly lenthening the cell, B F E D, as the grub grows older, and duly supply it with food, which they place before its mouth and round its body. The animal, which can only move in a spiral direction, keeps turning to take the jelly deposited before it, and thus slowly working downwards, arrives insensibly nearer the orifice of the cell, just at the time that it is ready to be metamorphosed into a nymph. At this moment, the workers, conscious of the impending change, seal up the mouth E F of the cell, and cease their attentions, leaving nature to effect the last transformation.

One of these cells is shown at *d*, in fig. 49.

That the mere change in the quality of the food, combined with the increased capacity and altered form of the cradle, should be the means of producing a transformation, so extreme as that from a worker to a queen, must be a matter of profound astonishment to every reflecting mind ; so much so indeed, that without the most incontestable evidence, and the power moreover of reproducing the phenomenon at will, it could not be credited. Let any one imagine how such an assertion as this, that the foal of an ass by a particular sort of provender, and by being reared in a stable of particular magnitude and form, could be made to grow into a through bred horse, would be received. Yet, such a transformation produced by such means would not be one whit more wonderful than the change of a worker grub into a queen-bee, by the means just stated. "What!" says Kirby, addressing his correspondent, "you will ask, can a larger and warmer house, a different and more pungent food, and a vertical instead of an horizontal posture, give a bee a different-shaped tongue and mandibles ; render the surface of its under-legs flat instead of concave ; deprive them of the fringe of hairs that forms the basket for carrying the masses of pollen,—of the auricle and pecten which enable the workers to use these legs or feet as pincers,—of the brush that lines the insides of the feet ? Can they lengthen its abdomen ; alter its colour and clothing ; give a curvature to its sting ; deprive it of its wax pockets ; and of the vessels for secreting that substance ; and render its ovaries more conspicuous and capable of yielding worker and drone eggs ? "

In the next place, can the apparently trivial circumstances just mentioned alter altogether the instincts of these creatures ? Can they give to one description of animals address and industry, and to the other astonishing fecundity ? Can we conceive them to change their very passions, tempers, and manners ? That the very same fœtus, if fed with more pungent food, in a higher temperature, and in a vertical position, shall become a female, destined to enjoy love, to burn with jealousy and anger, to be incited to vengeance, and to pass her time without labour—that this very same fœtus, if fed with more simple food, in a lower temperature, in a more confined and horizontal habitation, shall come forth a worker, zealous for the good of the community, a defender of the public rights, enjoying an immunity from the stimulus of sexual appetite and the pains of parturition—laborious, industrious, patient, ingenious, skilful,—incessantly engaged in the nurture of the young, in collecting honey and pollen; in elaborating wax; in constructing cells, and the like; paying the most respectful and assiduous attention to objects which, had its ovaries been developed, it would have hated and pursued with the most vindictive fury until it had destroyed them! Further, that these factitious queens, thus produced from worker eggs treated as above described, shall differ remarkably from the natural queens proceeding from royal eggs in being altogether mute! All this must seem so improbable, and next to impossible, that it would require the strongest and most irrefragable evidence to establish it.*

141. It will be remembered that the princesses, when forcibly confined to their native cells by the workers on guard over them, after they have undergone the last transformation, utter a peculiar sound, to the varieties of which Huber ascribes the power of the workers to determine their relative ages. Kirby in the observations just quoted, refers to this, when he indicates one of the distinctions between the factitious and natural queens, the former never uttering these or any other sounds.

142. Another remarkable distinction between the factitious and natural queens is indicated by Huber; no guard is kept at the doors of the cells of factitious princesses, like that which has been already described in the case of the cells of natural princesses. The factitious princesses, unlike the natural, are not detained in their cells after they have undergone the last transformation, but are allowed to issue forth, if they have not been already destroyed by the jealous rage of the first which comes to life.

This peculiarity in the policy of the hive may be explained by the fact, that while the natural princesses are wanted to take the sovereignties of the successive swarms, the factitious ones are only produced to meet the extraordinary emergency of the hive being deprived of its queen, leaving behind her no royal brood, and since only one queen is wanted, the factitious princesses are allowed, and indeed encouraged, by the workers to engage in

* Kirby, Int., vol. ii. 110.·

martial conflict until one only survives, who assumes the throne of the hive.

143. The circumstances and anecdotes related by observers illustrative of the affection, devotion, and respect manifested towards the queen by her subjects are innumerable. In addition to those which we have already given, the following will be read with interest.

All the devotion, it must be observed, commences only after the royal nuptials. A virgin queen is treated with indifference the most absolute. But after her marriage has been celebrated, and she presents herself to her subjects in the double character of sovereign and mother, they more than respect her. "They are," says Reaumur,[*] "constantly on the watch to make themselves useful to her, and to render her every kind office. They are for ever offering her honey. They lick her with their proboscis, and wherever she goes she has a court to attend her."

144. The same naturalist relates that even the inanimate body of the queen is an object of tenderness and affection to the bees. He took one out of the water quite motionless and seemingly dead. It was also mutilated, having lost part of one of its legs. Bringing it home, he placed it among some workers that he had found in the same situation, most of which he had recovered by means of warmth, some, however, being still in as bad a state as the poor queen. No sooner did these revived workers perceive the latter in this wretched condition than they appeared to compassionate her case, and did not cease to lick her with their tongues till she showed signs of returning animation; which the bees no sooner perceived than they set up a general hum as if for joy at the happy event. All this time they paid no attention to the workers, who were in a most miserable condition.[†]

145. In the economy of the bee, there is nothing which presents more difficulty to the naturalist than the satisfactory explanation of the functions of the drones. These, as has been already explained, are the sole male members of the society; the queen being the sole fertile female; and the workers, though female, exercising none of the functions of that sex, and being limited to the industrial and parental duties of the society. The number of drones in a single society is from 1500 to 2000, one only of whom can enjoy the honour of elevation to the distinguished position of king consort, and that one, as already explained, never surviving the day of the nuptials. What then, it may well be asked, are the services rendered to the community by these hundreds of consumers of the products of the industry of the society? They never themselves take part in the common labours. They neither

* Reaumur, v. 262.   † Reaumur, v. 265.

collect food nor materials, nor do they aid in any way in the construction of the dwellings, nor in the care or nurture of the young. In the absence of any better explanation of their vast number it has been said that the purpose is to insure a consort to the queen. But surely this object might be effected without encumbering the society with 2000 candidates for the royal favour.

It has been suggested by some apiarists that the drones may sit upon the eggs, and by others that their use may be to develope heat sufficient to maintain the hive at the necessary temperature ; but the experiments and observations of other naturalists have set aside these hypotheses.

146. Whatever be the purpose which this section of the society is destined to fulfil, their treatment by the people, and the manner in which their existence is terminated, are remarkable.

So long as swarms continue to issue from the hive, drones are wanted to supply the necessary proportion of that class to accompany them. But after the swarming season closes, which in these climates it generally does towards the end of July, at least in dry summers, the general massacre of the drones takes place. At that time the bees are seen hunting them in all parts of the hive, and driving them to the base upon which it stands. Soon after this the stand and the ground before the hive are found to be covered with the bodies of hundreds of the murdered drones. It was supposed by Bonnet that no direct massacre was executed, but that the drones driven from the stores of their food died of starvation.*

147. Huber, however, among his other numerous discoveries, contrived to witness, through the eyes of his faithful Burnens, the actual massacre.

At the season at which the extermination usually took place, he placed upon plates of glass six populous hives occupied by swarms of the preceding year, and Burnens lying on his back under the hives was enabled to witness all that took place by the transparency of their bases. On the 4th of July, 1787, he witnessed the massacre, which took place at the same hour in all the six hives. The base was crowded with bees, who appeared in a state of great excitement. As fast as the drones, hunted by other bees from the superior parts of the combs, arrived at the base, the bees there assembled fell upon them, seizing them by their antennæ, legs, or wings, and after dragging them about with apparent rage, put them to death by stabbing them with their stings between the segments of the abdomen. The moment they were thus pierced, they spread their wings and expired. However,

* Bonnet, "Contemplation de la Nature," chap. xxvi. part. xi.

as if the workers did not feel sufficiently certain of their fate, they continued to pierce their bodies with their stings, and often drove these formidable weapons in so deep that they could only extricate them by unscrewing them in the manner already described (126).

The next day they resumed their observations, when a most curious spectacle presented itself. During three hours they saw the massacre of drones, which had been resumed with the same fury, continued. On the preceding day they had exterminated all the drones of their own hives; but this time their attack was directed against those of neighbouring hives, which, having fled, had taken refuge in these, after the massacre of the preceding day had been concluded.

Not content with this complete extermination of the drones themselves, the workers resorted to the cells in which drone nymphs were contained, which had not yet completed their final transformation. These they pitilessly dragged forth, killed, sucked the juices contained in their bodies, and then flung the carcasses out of the hives.

148. It was also ascertained by Huber, that in hives deprived of their queen, or in which the queen, by reason of retarded fecundation, only laid drone eggs, no massacre ever took place. In such hives the drones not only find a sure refuge, but are carefully nurtured and fed.

This circumstance, combined with the fact that the massacre never takes place until after the swarming season is over, seems to indicate the functions of the drones. They are useful only where candidates for the royal nuptials are likely to be wanted.

149. The most interesting class of the bee community is also that which is by far the most numerous, the workers. Indeed, to this class all others must be regarded as subordinate, just as in human societies all are dependent on the producing classes. Much respecting their character, habits, and manners, in relation to the care of their young, and the construction of the city, in a word in respect to their internal labours, has been already explained. Something now must be said of their external industry, directed to the collection of provisions for the community, young and old, and of the materials necessary for the prosecution of all their various works, labours which have been illustrated by Professor Smyth in the following beautiful lines :—

> " Thou cheerful bee ! come, freely come,
> And travel round my woodbine bower ;
> Delight me with thy wandering hum,
> And rouse me from my musing hour.

Oh ! try no more those tedious fields,
Come taste the sweets my garden yields ;
The treasures of each blooming mine,
The bud, the blossom—all are thine.

" And, careless of this noontide heat,
I'll follow as thy ramble guides ;
To watch thee pause and chafe thy feet,
And sweep them o'er thy downy sides ;
Then in a flower's bell nestling lie,
And all thy envied ardour ply !
And o'er the stem, though fair it grow,
With touch rejecting, glance and go.

" Oh, Nature kind ! Oh, labourer wise !
That roam'st along the summer's ray,
Glean'st every bliss thy life supplies,
And meet'st prepared thy winter day !
Go, envied, go—with crowded gates
The hive thy rich return awaits ;
Bear home thy store, in triumph gay,
And shame each idler of the day."

150. The immediate objects to which the exterior industry of the bee is directed, are *nectar, pollen,* and *propolis.*

Nectar is a specific juice, found in certain classes of flowers, from which the bee elaborates honey and wax.

Pollen is a peculiar powder, or dust, spread over the anthers of flowers, which constitutes the principle of fecundation of the flowers themselves, and is the material of which the bee makes bread, which serves as food both for old and young.

Propolis is a resinous substance, evolved by certain vegetables which the bee uses as cement, mortar, or glue, in its architecture. When the bee pierces the vessels of the flowers, which, containing nectar, are called nectarines, and swallows that precious juice, it is deposited provisionally in the honey-bag already described (26) ; sometimes called, on that account, the first stomach. Here this nectar is converted into honey, the chief part of which is regurgitated, to be stored up for future general consumption in the honey-cells of the combs.

In the stomach, properly so called (26), and in the intestines, the bread only is found.

How the wax is secreted, physiologists have not yet discovered with any certainty. It is evident, however, that the immediate seat of its production is within the abdomen, since the parts called wax-pockets, from which it is externally evolved, are rendered visible by pressing the abdomen so as to make it extend itself. A pair of quadrangular whitish pockets, of soft membranaceous texture, will then be seen on each of the four middle ventral, segments. On these the plates of wax are formed, and are found upon them in different states so as to be more or less perceptible.

151. Observe a bee, says Kirby, that has alighted on a flower. The hum produced by the motions of her wings ceases, and her work begins. In an instant she unfolds her tongue, which was previously rolled up under her head. With what rapidity does she dart this organ between the petals and the stamina! At one time she extends it to its full length, then she contracts it; she moves it about in all directions, so that it may be applied to the concave and convex surface of the petal, and sweep them both, and thus by a virtuous theft, she robs it of all its nectar. All the while this is going on, she keeps herself in a state of constant vibratory motion.

Flowers, though the chief, are not the only sources from which the bee derives the material of honey and wax. She will also eat sugar in every form, treacle, the juice secreted by aphides; and, in fine, the juice of the bodies of nymphs and of eggs of bees themselves, as already explained.

152. When the industrious little creature has filled its honey-bag with nectar, it proceeds to collect the pollen, of which it robs the flowers by brushing it off with the feathery hairs with which its body is covered. As the honey is called the NECTAR, so this pollen, or the substance bee-bread, into which it is converted, may be called the AMBROSIA of the hive. Together they constitute the food and the drink of the population.

When the bee has so rolled itself in this farina of the blossoms of the garden and the field, that its whole body is so powdered with it, as to give it the peculiar colour of the species of flowers to which it happens to resort, it suspends its excursions, and sets about to brush its body with its legs, which, as already explained, are supplied with brushes for this express purpose. Every particle of the flower thus brushed off is most carefully collected and kneaded up into two little masses, which are transferred from the fore to the hind legs, and there packed up into the baskets provided for its reception and transportation.

Naturalists generally are of opinion that in each of its excursions a bee confines its foraging operations to a single species of flower. This explains the fact that the colour of their load after such excursions is uniform, depending on the particular species of flower which they have robbed of its sweets. Thus, according to Reaumur, some bees are observed to return loaded with red pellets on their thighs, others with yellow, others whitish, and others with green.

Kirby observes, that it seems probable that the bee confines its operations in such excursions to flowers of the same species, and that the grains of pollen which enter into the same mass should be homogeneous, and consequently fitted by their physical properties to cohere with greater facility and firmness.

153. But connected with this, another important purpose of nature is fulfilled, which must not here pass without special notice. The principle, so fruitful in important social consequences among animals, that the offspring owes its parentage jointly to two individuals of different sexes, or, in other words, must always have a father and a mother, equally prevails in the vegetable kingdom. There also are the gentlemen and ladies, there also are the loves which unite them, loves which as well as those of superior orders of beings have supplied a theme for poets.* Now among the many other interesting offices with which the Author of nature has invested the little creatures, which form the subject of this notice, not the least singular is that of being the priests who celebrate the nuptials of the flowers. It is the bee literally which joins the hands and consecrates the union of the fair virgin lily and the blushing maiden rose with their respective bridegrooms. The grains of pollen which we have been describing are these brides and bridegrooms, and are transported on the bee from the male to the female flower; the happy individuals thus united in the bands of wedlock being the particular grains, which the bee lets fall from its body on the flower of the opposite sex, as it passes through its blossom.

154. And here we find another circumstance to excite our admiration of the wise laws of that Providence, which cares for the well-being of a little flower, as much as for that of a great lord of the creation. If the bee wandered indifferently from flower to flower without selection, the gentlemen of one species would be mated with the ladies of another, hybrid breeds would ensue, and the confusion of species would be the consequence. But the bee, as knowing this, flies from rose to rose, or from lily to lily, but never from the lily to the rose, or from the rose to the lily.

155. When a bee, laden with pollen, arrives in the hive, she sometimes stops at the entrance, and leisurely detaching it piecemeal from her legs, devours it bit by bit. Sometimes she passes into the hive and walks over the combs, or stands stationary upon them, but whether moving or standing never ceases flapping her wings. The noise thus produced, a sort of buzzing, seems to be a call understood by the populace within hearing, for three or four of them immediately approach and surround her. They begin to aid her to disembarrass herself of her load, each taking and swallowing more or less of her ambrosia until the whole is disposed of.

* Darwin's Loves of the Plants.

156. When more pollen has been collected than the society wants for present use, it is stored up in some of the unoccupied cells. The bee, laden with it, puts her two hind legs into the cell, and with the intermediate pair pushes off the pellets. When this is done she, or another bee if she be too much fatigued, enters the cell head-foremost and remains there for some time, during which she is occupied in diluting, kneading, and packing the bee-bread; and so they proceed one after another, until the cell has been well packed and filled with the store of provisions. In some combs a large portion of the cells is filled with this ambrosia, in others, cells containing it are intermixed with those filled with honey or with bread. It is thus everywhere at hand for use.*

The propolis, the third object of bee industry, is collected from various trees, and especially from certain species of the poplar. It is soft and red, will allow of being drawn out into a thread, is aromatic, and imparts a gold-colour to white polished metals. It is employed in the hive, as already stated, not only in finishing the combs, but also in stopping up every chink and orifice by which cold, wet, or any enemy could enter. They coat with it the chief part of the inner surface of the hive, including that of the sticks placed there for the support of the comb. It is carried by the bees in the same manner as is the pollen on the hind legs.

157. The radius around their habitation, within which the bee industry is confined, is differently estimated, being according to some a mile, and according to others extending to a mile and a half. Various experiments prove that it is by their scent that the bees are guided to the localities where their favourite flowers abound.

* Kirby, Int., ii. 151.

Fig. 63.—Scotch hive.

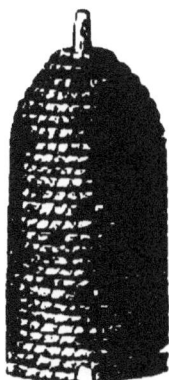

Fig. 64.—Radouau's hive.

Fig. 65.—Cork hive (South of France).

THE BEE.

CHAPTER VI.

158. How they fly straight back to the hive—manner of discovering the nests of wild bees in New England.—159. Average number of daily excursions.—160. Bee pasturage—transported to follow it—in Egypt and Greece.—161. Neatness of the bee.—162. Its enemies.—163. Death's-head moth.—164. Measures of defence adopted by Huber.—165. Measures adopted by the bees.—166. Wars between different hives.—167. Demolition of the defensive works when not needed.—168. Senses of insects.—169. Senses of the bee.—170. Smell.—171. Experiments of Huber.—172. Remarkable tenacity of memory.—173. Experiments to ascertain the organ of smell.—174. Repugnancy of the bee for its own poison.—175. Their method of ventilating the hive.—176. Their antipathy against certain persons.—177. Against red and black-haired persons.—178. Difference of opinion as to the functions of the antennæ.—179. Organs of taste.—180. Hearing: curious anecdotes.—181. Vision.—182. Peculiar characters of queens; royal old maid.—183. Drone-bearing queens.—184. Change of their instincts and manners.—185. Their treatment by the workers.—186. Nuptials never celebrated in the hive.—187. Effect of amputating the royal antennæ.

158. ONE of the many wonders presented by their economy is the directness and unerring certainty of their flight. While collecting their sweets they fly hither and thither, forward or backward, and right or left, as this or that blossom attracts them; but when fully laden with the spoil, though upwards of a mile from their city, they start for it in a course more exact than if they were guided

by a rudder and compass, governed by the hand of the most consummate navigator. By what means this is accomplished has never been explained, but connected with it is an account given in the "Philosophical Transactions" which we cannot refrain from quoting here. "In New England a species of wild hive-bees abounded in the forests about the year 1720. The following was the method practised for discovering their nests and obtaining their honey. The honey-hunters set a plate containing honey or sugar, upon the ground on a clear day. The bees soon discovered and attacked it. Having captured two or three who had thus gorged themselves, the hunter liberated one of them and marked the direction in which it flew. He then changed his position, walking in a direction at right angles to the course of the bee to a distance of a few hundred feet, where he liberated another of his little captives, and noted as before the direction of its flight. The point where the two directions thus obtained, intersected, was of course that to which both bees had directed their course, and there the nest was always found."

159. The industry of the bee may be estimated by the average number of its daily excursions from the hive to collect provisions. According to Reaumur, if the total number of excursions be divided by the total number of bees in a hive, the average number daily made by each bee would be from five to six. But as one-half of the bees are occupied exclusively with the domestic business of the society, either in nursing and tending the young, packing and storing the provisions, or constructing the combs, the total number of excursions must be divided, not between the whole, but between only half the total number of bees, which would give ten excursions to each individual of the collecting class; and if the average length of each excursion be taken at three quarters of a mile, this would give the average distance travelled by each collector as fifteen miles! It is estimated by Kirby that the quantity of ponderable matter thus transported during a season in a single hive would be about 100 lbs. "What a wonderful idea does this give of the industry and activity of those useful little creatures! and what a lesson do they read to the members of societies, that have both reason and religion to guide their exertions for the common good! Adorable is that Great Being who has gifted them with instincts which render them as instructive to us, if we will condescend to listen to them, as they are profitable." *

160. The plants and flowers which form the pasturage of the bees are, in many countries, produced at different places at different seasons of the year; and where the bees in a particular neigh-

* Kirby, Int., ii. 155.

bourhood are numerous, the pasturage surrounding their hives often becomes exhausted. In such cases the agriculturists transport the bees from localities which they have exhausted, to others in a state of comparative abundance, just as the shepherd drives his sheep from field to field, according as the pasturage is eaten down. In Egypt, towards the end of October, when the inundations of the Nile have ceased, and the husbandmen can sow the land, saintfoin is one of the first things sown; and as Upper is warmer than Lower Egypt, the saintfoin gets there first into flower. At this time bee-hives are transported in boats from all parts of Egypt into the upper district, and are there heaped in pyramids upon the boats prepared to receive them, each being marked with a number which indicates its owner. In this station they remain for some days, and when it is considered that they have pretty well exhausted the surrounding fields of their sweets, they are removed a few leagues lower down, where they are retained for a like interval; and so they descend the river, until towards the middle of February they arrive at its mouth, where they are distributed among their respective proprietors.*

A similar practice prevails in various parts of the East and in Greece. The inhabitants of the towns are often the proprietors of fifty or sixty hives, the product of which forms an article of their trade. The hives are sent in the season when the herbage is in flower to the various rural districts, being sealed up by the owner, the small bee-door only being open, and are given in charge to the villagers, who at the close of the season are paid for their care of them. Ranges, consisting of five or six hundred hives, are often seen thus put out to grass.†

161. Bees are remarkable for neatness and cleanliness, both as to their habitations and their persons. They remove all dirt and nuisances from their hive, with the regularity of the neatest housewives. When their strength is insufficient for this, they contrive various ingenious expedients to abate the nuisance. If snails find their way into the hive, as they sometimes do, they kill them with their stings; and in order to prevent noisome and unwholesome effluvia from their decomposing remains, they embalm them with propolis. If the snail is protected from their stings by its shell, they bury it alive in a mass of propolis.

When pressed by natural wants, they do not defile their habitation by relieving themselves in it, but go abroad for the purpose.

When a young bee issues from the cell, a worker immediately approaches, and, taking out its envelope, carries it out of the hive; another removes the exuviæ of the larva, and a third any

* Reaumur, v. 698.
† Willock, in "Gardeners' Chronicle, 1841, p. 84.

filth or ordure that may remain, or any pieces of wax that may have fallen in when the young bee broke through its cocoon. But they never attempt to remove the silk lining of the cell spun by the larva in its first transformation, because that, instead of being a nuisance, gives increased solidity and ornament to the cell.

162. Notwithstanding the amiable character and excellent political organisation of the bees, these little people have numerous enemies, with some of whom they are often compelled to wage offensive wars, and against others to fortify themselves, by expedients and with skill, which will bear comparison with the operations of the most consummate military engineers. Sebastopol itself was not more ingeniously defended by its outworks than, in certain cases, bee-hives are.

From the curious account which Latreille has given us of *Philanthus aviporus*, a wasp-like insect, it appears that great havoc is made by it of the unsuspecting workers, which it seizes while intent upon their daily labours, and carries off to feed its young.

163. Another insect, which one would not have suspected of marauding propensities, must here be introduced. Kuhn informs us, that long ago (in 1799) some monks who kept bees, observing that they made an unusual noise, lifted up the hive, when an animal flew out, which, to their great surprise, no doubt, for they at first took it for a bat, proved to be the death's-head hawk-moth (*Acherontia atropos*), already celebrated as the innocent cause of alarm; and he remembers that several, some years before, had been found dead in the bee-houses. M. Huber also, in 1804, discovered that it had made its way into his hives and those of his vicinity, and had robbed them of their honey. In Africa, we are told, it has the same propensity; which the Hottentots observing, in order to monopolise the honey of the wild bees, have persuaded the colonists that it inflicts a mortal wound.

This moth has the faculty of emitting a remarkable sound, which he supposes may produce an effect upon the bees of a hive, somewhat similar to that caused by the voice of their queen, which as soon as uttered strikes them motionless, and thus it may be enabled to commit with impunity such devastation in the midst of myriads of armed bands.

The larvæ of two species of moth (*Galleria cereana* and *Mellonella*) exhibit equal hardihood with equal impunity. They, indeed, pass the whole of their initiatory state in the midst of combs. Yet, in spite of the sting of the bees of a whole republic, they continue their depredations unmolested, sheltering themselves in tubes made of grains of wax, and lined with silken tapestry, spun and woven by themselves, which the bees (however disposed they may be to revenge the mischief which they do to them, by devouring what to all other animals would be indigestible—their wax) are unable to penetrate. These larvæ are sometimes so numerous in a hive, and commit such extensive ravages, as to force the poor bees to desert it and seek another habitation." *

164. Huber gives the following most interesting account of the measures taken by his bees, to fortify themselves against the incursions of the death's-head moth.

When he found his hives attacked and their store of honey pillaged by these depredators, he contracted the opening left for the exit and entrance of the bees to such an extent, as while it allowed them free ingress and egress, it was so small that their plunderers could not pass through it. This was found to be perfectly effectual, and all pillage was thenceforward discontinued in the hives thus protected.

165. But it happened that in some of the hives this precaution was not adopted, and here the most wonderful proceeding on the part of the bees took place. Human contrivance was brought into immediate juxtaposition with apiarian ingenuity.

The bees of the undefended hives raised a wall across the gate of their city, consisting of a stiff cement made of wax and propolis mixed in a certain proportion. This wall, sometimes carried directly across and sometimes a little behind the door, first completely closed up the entrance; but they pierced in it some openings just large enough to allow two bees to pass each other in their exits and entrances.

The little engineers did not follow one invariable plan in these defensive works, but modified them according to circumstances. In some cases a single wall, having small wickets worked through it at certain points, was constructed. In others several walls were erected one within the other, placed parallel to each other, with trenches between them wide enough to allow two bees to pass each other. In each of these parallel walls several openings or wickets were pierced, but so placed as not to correspond in position, so that in entering a bee would have to follow a zigzag course in passing from wicket to wicket. In some cases these walls or curtains were wrought into a series of arcades, but so that the intervening columns of one corresponded to the arcades of the other.

The bees never constructed these works of defence without urgent necessity. Thus, in seasons or in localities where the death's-head moth did not prevail, no such expedients were resorted to. Nor were they used against enemies which were open to attack by their sting. The bee, therefore, understands

* Kirby, vol. i. p. 130.

not merely the art of offensive war, and can play the part of the common soldier, but is also a consummate military engineer; and it is not against the death's-head moth alone that it shows itself capable of erecting such defences.

166. Thinly peopled hives are sometimes attacked by the population of other bee cities. In such cases, incapable of immediate defence by reason of their inferior numbers, they erect similar fortifications, but in this case they make the wickets in the walls so small that a single worker only can pass through them; and a small number stationed on the inside of these openings, are accordingly sufficient to defend the hive against the attack of large besieging armies.

167. But when the season for swarming arrived, these works of defence, whether constructed against the invasion of the moth or hostile bees, became an impracticable obstruction to the exit of the succession of emigrating colonies, and were therefore demolished, and were not reconstructed without pressing necessity. Thus the works constructed in 1804 against the invasions of the moth were taken down in the swarming season of 1805; and as the plunderers did not re-appear in that year, they were not re-erected. But in the autumn of 1807, the moths appearing in great numbers, the bees immediately erected strong barricades, and thus effectually prevented the disaster with which their population was menaced. In the next swarming season, in May 1808, these works were again demolished.

It ought to be observed, that whenever the door of the hive is itself too small to admit the moth, the bees erect no defences against it.*

168. One of the most interesting and, at the same time, most difficult question connected with the faculties of insects, is that of the number and nature of their senses. It has been often and truly said, that no being, however intelligent, can form even the most obscure notion of a sense of which he is himself deprived. The man deprived of sight, to whom the colour scarlet was elaborately described, said that his notion of it was that of the sound of a trumpet. Granting then the possibility that insects may be endowed with a peculiar sense, or mode of perception, of which we are destitute, we are in no condition to form a conception of the power or impressions of such a sense, any more than the blind man was who attempted to acquire a conception of a red colour.

But without supposing the possible existence of peculiar senses independent of the five with which we are endowed, it may be that the very organs which we possess may be given with an infi-

* Huber, ii. 293—298.

nitely higher degree of sensibility to these minute species. Their auditory organs may be such as to give them the power of ear-trumpets, and their eyes may be either microscopic or telescopic, or both united. Their olfactory organs may have a susceptibility infinitely more exalted than ours, as indeed innumerable facts prove those of many species of inferior animals to be. Art and science have supplied us with numerous tests, by which the physical properties of substances are distinguished, by characters which escape all our senses. Why may not the Creator have given to inferior animals specific organs, capable of perceiving those distinctions, as surely and promptly as the eye distinguishes shades of colour, the nose varieties of odour, or the ear the pitch of a musical note?

169. Among social insects, the hive-bee stands preeminent for the manifestation of sensitive faculties. Sight, touch, smell, and taste, are universally accorded to it. Hearing was regarded as doubtful, but we have shown that a noise produced at any side of a hive, will immediately bring there the queen and her court, to see what is the matter.

But if the sensibility of the ear be doubted, what exaltation of power do we not find in the eye! How unerring is the perception of her dwelling, while the bee lies at distances and under circumstances, which might well appear to baffle the most acute human organ, aided even by human intelligence! The little bee, issuing from her hive, departs upon her industrial excursion, and flies straight to the field which she has already discovered to be most fertile of honey flowers. Her route to it is as straight as the flight of a bullet from a gun to the object aimed at. When she has gathered her load, she rises in the air, and, flying back to her hive with the same unerring certainty, finds it among many, and entering it, finds the cells which are appropriated to her care.

The sense of touch is, perhaps, even more to be admired than that of sight, for it supplies the place of that sense in the darkness of the internal labyrinth of the hive. In darkness the architecture of the combs is constructed, the honey is stored in the cells appropriated to it, the young are nourished, their food being varied with their respective ages, the queen is recognised,—and all this appears to be accomplished by some sensitive power possessed by the antennæ, organs whose structure, nevertheless, seems to be incomparably inferior to that of the human hands.

The industrial activity of the bee is much less excited by warm weather and bright sunshine, than by the prospect of collecting an abundant supply of provisions for the hive. When the lindens and the buck-wheat are in flower, they brave the rain and cold, commencing their excursions before sunrise, and continuing their work much later than their customary hours. But when the flowers rich in pollen and nectar prevail in less abundance, and when the scythe has swept away the flowers which enamelled the fields, even the brightest sunshine and the warmest days fail to attract the industrious population to go abroad.

170. Of all the senses of the bee, that of smell appears to be the most acute. Certain odours have an irresistible attraction for the insect, while others are in the same degree repugnant to it. Of the former, as might naturally be expected, honey is by far the most exciting. It was supposed by Huber, not without much probability, that the bee is attracted to this or that flower, not by its colour, form, or other visible properties, but by the odour of the nectar it contains. To test this experimentally, Huber put some honey in a box, so as to be invisible from the outside, and placing it in the neighbourhood of his hives, found that the bees crowded round it in a few minutes, finding their way to the honey through a small hole left for the purpose.

171. He next made several small entrance holes in a box containing honey, but covered each hole with a sort of card valve, such that it would be possible for a bee to raise it and enter the box. The box thus prepared was placed at two hundred yards from the hives. In half an hour the bees found it, crowded in great numbers on every side of it, examining carefully every part, as if to seek for an entrance. At length, finding the valves, they set to work at them, and never ceased until they succeeded in raising them, when they entered and took possession of the spoil.

How exquisitely acute must be their olfactory organs will be apparent, when it is considered that, in this case, the box and valves must have confined very nearly the whole effluvia of the honey.

172. The following remarkable proof of the tenacity of memory with which the bee is endowed, is given by Huber. A supply of honey had been placed in autumn upon an open window. The bees had the habit of coming to feast upon it. This honey being removed, the window was closed, and remained closed during the winter. In the following spring the bees again found their way to the same window, expecting again to find a supply there, although none had been placed there. It is evident in this case, that the insect must have been guided by its memory alone, and that it was capable of retaining a recollection of places and circumstances for several months.

173. Huber made several curious and interesting experiments to determine the seat of the sense of smell. If, as was natural to expect, it were situate in some of the appendages of the mouth,

it would be deadened by stopping these, as we defend ourselves from a noisome odour by stopping the nose. Catching several bees he, therefore, held them while he stopped their mouths and probosces with flour-paste, and liberating them when the paste was hardened, he found that they no longer showed any sign of the possession of a sense of smell. They were neither attracted by honey, nor repelled by objects whose odours were known to be most repugnant to them.

174. Among the substances to whose odour the bee shows the strongest repugnance, is its own poison. This was demonstrated by Huber by very remarkable experiments. Having provoked the insect to put forth its sting, and eject its poison, he presented this offensive juice on the end of a sharp instrument to some worker bees, which were quietly resting at the door of their hive. A general agitation was immediately manifested among them. Some launched themselves on the poisoned instrument, and others fell upon the individual who held it. That it was not the instrument itself which in this case provoked their rage, was proved by the fact, that a similar one, bearing no poison, being presented to them, did not produce any effect.

175. An inconvenient elevation of temperature and want of ventilation will sometimes impel the bees to leave their combs, but if they are excited to remain upon them by the want of feeding, they know how to reconcile the conflicting impulses. In that case they produce coolness and change of air without deserting the provisions which surround them, or the care of their young. A certain number of the insects begin to flap their wings, which are thus used as fans, producing currents of air. But as they are not able to sustain this labour for an indefinite time, they take it by turns, regularly relieving each other.

To try what the conduct of the bees would be, if by artificial means the ventilation of the hive were so impeded that the usual small number of *fanners* would not suffice, Huber submitted hives to such unusual conditions, and found that in such cases the number of bees flapping their wings was augmented in the same proportion as the ventilation was impeded, until at length the whole population of the hive were thus occupied.

176. The antipathy which bees manifest against particular individuals, is generally ascribed to some odour proceeding from their persons to which the insect bears a repugnance. M. de Hafor, of the Grand Duchy of Baden, had been for many years an assiduous cultivator and amateur of bees, and was on such friendly terms with them that he could at all times approach them with impunity. He would, for example, put his fingers among them, select the queen, and taking hold of her, place her on the palm of his

hand. It happened that this gentleman was attacked with a violent and malignant fever, which long confined him to his bed and his house. Upon his recovery he, naturally enough, revisited his old friends the bees, and began to caress them and renew his former familiarity.

He found, however, to his surprise and disappointment, that he was no longer in possession of their favour, and instead of being received as formerly, his advances were resented as an unwelcome and irksome intrusion; nor was he ever afterwards able to perform any of the usual operations upon them, or to approach them without exciting their rage.

177. According to Dr. Bevan and M. Feburier, both close and accurate observers of the habits of the insect, red and black-haired persons are peculiarly obnoxious to it. Feburier mentions a mastiff to which his bees had a particular aversion, pursuing him into the house with such pertinacity, that doors and windows were obliged to be closed for his protection.

Dr. Bevan mentions that he had two friends, brothers, one of whom was so inoffensive to the bees, that he could stand with impunity over the hive and watch all their doings, while the other could scarcely enter the garden with impunity.

178. The antennæ are generally regarded as the proper organs of the tactile sense, and hence are popularly, though not properly, called feelers,—the feelers being in fact the palpi already mentioned. Naturalists are not agreed as to the functions of the antennæ, though all concur as to their importance. Some consider them as organs of smell, others as organs of hearing; while others claim for them the place of organs of a sixth sense, of which man and the higher animals are destitute. This sense is considered by Kirby as an intermediate faculty between sight and hearing, rendering the insect sensible of the slightest movement of the circumambient air. Dr. Evans, as quoted by Dr. Bevan, in reference to the faculty conferred on the bee by the antennæ, says,—

> " The same keen horns, within the dark abode,
> Trace for the sightless throng a ready road ;
> While all the mazy threads of touch convey
> That inward to the mind, a semblant day."

The antennæ, and the two pair of palpi, would seem to have correlative and complementary functions : they are both in constant motion. The palpi are in reality the feelers, in the proper sense of the term; as is apparent by observing the manner in which the insect applies them to the food before eating it.

179. Cuvier considers the organs of taste in the bee to constitute one of its most important characters. The sensibility of these

organs is manifested by the delicate choice of food which the insect makes, showing a preference for those flowers, wherever they can be found, which yield the finest honey. Hence the celebrity of the honey of Narbonne, Hymettus, Hybla, and Pontus.

180. Numerous indications show that the bee possesses the sense of hearing. The manner in which they are attracted to any quarter of the hive where an unusual noise is produced, has been already mentioned. Dr. Bevan mentions some curious examples of their power of hearing, and even of the sense they seem to attach to particular vocal sounds. Thus he mentions an old dame of his acquaintance, who was a very fearless operator in the treatment of these insects, and who used to suppress any movement of anger on the part of the bees merely by saying to them, "Ah! would you dare?" A servant of Mr. Knight, the well-known apiarian, used to quell their anger by exclaiming, "Get along, you little fools!"

Some difference of opinion has nevertheless prevailed as to the existence of this sense in insects. The opinion of Linnæus and Bonnet was against it. Many evidences, however, may be adduced in favour of its existence. Thus, one grasshopper will chirp in response to another, and the female will be attracted by the voice of the male. Brunelli shut up a male in a box, and allowed the female her liberty; as soon as the male chirped she flew to him immediately. A bee on the window within a bee-house will make a responsive buzz to its fellows on the outside.*

181. The indications of a keen sense of vision, in the certainty and precision with which the bee flies to its pasturage and back to its hive, have been already mentioned. Naturalists, however, are not agreed as to the particular power of the eyes of these insects. Some, for example, contend that their sight is extremely short, and that

> Its feeble ray scarce spreads
> An inch around;

while others contend that its vision of near objects is obscure and imperfect, but for distant ones quite distinct. Thus Butler and Wildman say that they have observed the bees go up and down seeking the door of the hive, as if they were in the dark; but Bevan observed that they easily discovered it by rising on the wing, and thus throwing themselves at a greater distance from it.

182. Among the mysteries of the social economy of the bee, there is perhaps nothing more curious than the circumstances which, in certain cases, appear to affect the personal character of the

* Bevan, p. 362.

sovereign. We have already explained that there are certain periods in the life of the queen, during which she produces eggs of certain sorts,—at one period those only of workers, at another those only of drones. But if the epoch of her nuptials be postponed to a certain advanced period of life, at which, if we may be allowed the expression, she begins to approach the condition of an *old maid*, a singular change is found to have taken place in her constitution, in consequence of which she is no longer capable of having any but *male offspring*, in other words, she is incapable of laying any but drone eggs.

183. Now since such a queen is obviously incapable of discharging those functions, which are indispensable to the continuance of the population over which she presides, and of whose young she ought, in the ordinary course of nature, to be common mother, it might be inferred that the instincts of the insects would lead them to disembarrass themselves of a sovereign, incapable of discharging the most important functions of her office, and to substitute for her, as we know they always have the power to do, one who should enjoy the plenitude of these functions.

184. Among the innumerable experiments of Huber, those are not least interesting which were directed to this point, that is to say, to submit the faculties of the queen to tests supplied by artificial means, contrived for placing her in social conditions, in which it could scarcely ever ·happen that she should find herself in the common course of *bee-nature*.

The first question which suggested itself to the great naturalist, was to ascertain whether queens, who thus married so late in life as to have only drone offspring, would exhibit the same spirit of jealous hostility towards the tenants of royal cells, and the future aspirants to thrones, as is invariably manifested by younger royal brides. To determine this it was necessary to place such a queen in a queenless hive, in which, however, there was at least one royal cell tenanted by a princess. Huber, therefore, placed the queen, who had not married until she had bordered upon old maidenhood, in a hive which had no queen, but in which there was one royal cell occupied by a princess. The old bride, whose nuptials had not been celebrated until she had attained the twenty-eighth day of her age, laid nothing of course except drone eggs. On being placed in the hive she exhibited none of the usual signs of hostility against the royal cell. On the contrary, she passed and repassed it many times a day without seeming to take the least notice of it, or to distinguish it in any way from the numerous cells which surrounded it on every side. In such of these latter cells as were unoccupied she deposited eggs, and notwithstanding the jealous guard which the workers kept around the royal cell occupied by the princess, the queen did not appear either to show a disposition to attack the imprisoned princess, or to fear any attack on the part of the latter.

185. Meanwhile the workers exhibited towards the queen the same respect and homage, lavished upon her the same affectionate cares, offered her honey, and formed round her in the same respectful circle, as they are wont to do round a sovereign possessing all the functions necessary to perpetuate the race.

It appears, therefore, that the postponement of the royal nuptials beyond a certain age, while it deprives the queen of the faculty of having any but male offspring, also deprives her of that instinctive feeling of jealous hostility towards rival queens, which forms a trait so remarkable in the characters of queens, whose nuptials take place at an earlier and more natural age.

To those who regard these little creatures as mere pieces of mechanism, obeying unreflecting impulses, having purposes always directed to the fulfilment of some important end in their economy, it will doubtless be surprising that members of the community so useless as those princesses, who postpone their nuptials until they are incapable of bearing worthier offspring, should not be destroyed as the drones are, after they cease to be useful. So contrary to this, however, is the fact, that no royal bride, however young, is the object of solicitude more tender, affection more sincere, and homage more profound, than those drone-bearing mothers. " I have seen," says Huber, " the workers lavish the most tender care upon such a queen, and, after her decease, surround her inanimate body with the same respect and homage as they had paid to herself while living, and, in the presence of these beloved remains, refuse all attention to young and fertile queens who were offered to them." * It must be admitted that this looks much more like the tenderness of moral affection than the mechanical impression of blind instinct.

186. We have already stated that the royal nuptials are always celebrated in the air, and under the bright beams of the sun, where the bride rises with her numerous suitors, and makes her choice. This bridal excursion into the fields of ether is so intimately interwoven with the customs of these little people, that if, by cutting off her wings before her nuptials, her majesty is deprived of the power of flight, she is consigned irretrievably to a life of single blessedness, since she can never submit to nuptials celebrated in the recesses of the hive, instead of the gay and bright sunshine of the free air.

* It will be observed that, according to the general habit of the blind, Huber uses the language of vision, and describes what he saw with the eyes of Bernens as if he had seen them with his own.

Lest it might be imagined, as indeed Swammerdam supposed, that the marriage is really consummated in this case in the hive, and that her majesty is only rendered sterile by the mutilation she has undergone, Huber cut off the wings of a queen immediately after the royal nuptials, but before her majesty had yet any offspring. In this case, however, her fertility was as great as usual, and she produced the customary number and variety of eggs.

187. One of the questions in insect physiology, which has been attended with a certain degree of doubt, is that which regards the functions of the antennæ. Huber, therefore, desiring to ascertain how the queen would be affected by the privation of these organs, cut off first one and then both, observing the conduct of her majesty after such mutilation.

The excision of one only of the antennæ produced no discoverable effect upon her faculties or conduct, but the amputation of both was followed by some very remarkable consequences.

The antennæ of a queen of limited fertility, who was incapable of having other than drone offspring, were cut off. From the moment she lost these organs she appeared to be affected by a sort of delirious intoxication. She ran over the combs with extraordinary vivacity. She did not give her suite, who formed the usual circle around her, time to make way for her, but rushed madly through them, violently breaking their ranks. She did not deposit her eggs in cells, but dropped them at hazard. The hive not being very full, there were parts of it unoccupied by combs. To these parts she rushed, and remained there a considerable time quiescent, appearing to avoid the presence of her subjects. Some of them, nevertheless, followed her to these deserted places, and eagerly testified their solicitude for her, caressing her, and offering her honey. This she generally declined; and when now and then she seemed disposed reluctantly to accept it, she appeared to lose the power of presenting her proboscis to receive it, directing that organ at one time to the head and at another to the legs of the workers, so that it was only by chance it encountered their mouths. She would then run back to the combs, and from the combs to the glazed sides of the hive, in wild delirium, never ceasing to drop her eggs here and there as she went along.

At other moments she seemed to be tormented with a desire to quit the hive, and rushed to the door for that purpose, but the orifice being too small to allow her body to pass through it, she was forced to desist, and returned to the interior. Notwithstanding this state of delirium, the bees never ceased to lavish upon her those cares which they are accustomed to bestow on their queen; but she received them with indifference.

Whether all this singularity and eccentricity of conduct was to be ascribed to the excision of the antennæ, or to that mutilation combined with the partial sterility which limited her offspring to drones, was not clear. To decide this point, Huber amputated the antennæ of a perfect queen, married at an early age, and who was bearing a numerous offspring, consisting of workers, drones, and princesses. This queen he placed in the same hive with the former, with a view to determine at once two questions, the one relating to the general conduct of the amputated queen, and the other, that which regarded the mutual bearing of two mutilated personages.

The general conduct was the same as that of the former queen. There was the same wild delirium ; the same rushing here and there as if under the influence of intoxication ; the same efforts to escape from the hive; and, in a word, the same peculiarity of conduct and manners. A like difference was apparent in their conduct towards each other. Instead of entering into deadly combat, as queens in their natural state would have done in like circumstances, they met and passed each other again and again without the slightest indication of mutual hostility. This is perhaps the strongest proof which can be obtained, that the privation of the antennæ utterly subverted their natural instincts.

Another curious social anomaly was manifested on this occasion. It will be recollected that where a strange queen is introduced into a hive over which a regular sovereign already presides, the population surround her, confine her as a prisoner within a ring of sentinels, and refuse to permit her to enter their city. In the present case, no such measures were adopted. On the contrary, the second mutilated queen was received with the same signs of welcome, and immediately became the same object of attention and homage as the first.

But the most wonderful fact of all those developed in this series of experiments, was that when a third queen in the perfect state, without mutilation, was introduced, the bees who had already treated the other two so well, immediately proceeded to maltreat this third and perfect queen. They seized her, dragged her about, bit her, and so closely surrounded her as to leave her room neither to move nor to breathe.

Having observed the apparent desire of these mutilated queens to issue from the hive, which they were only prevented from doing by the limited magnitude of the door, and desiring to see whether the bees or any considerable number of them would depart with her, as they would do with a perfect queen, Huber, after taking away the two queens who were sterile, or partially so, and leaving her who was fruitful in all respects, but deprived of her antennæ, he enlarged the door so as to allow her free passage through it. So soon as this was done, she went out, and took flight, but not a single bee accompanied her. She was, moreover, so heavy, being full of eggs, that she was not able long to sustain herself on the wing, and fell to the ground.

Various conjectures are made by Huber to explain this singular departure from the prevailing habits of the insect, but none of them appear so satisfactory as to require to be reproduced.

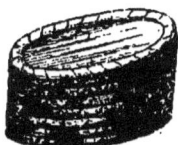

Fig. 85.—Oblique piece to elevate a village hive.

Fig. 86.—The bee-dress.

THE BEE.

CHAPTER VII.

188. Apiculture.—189. Suitable localities and pasturage.—190. The Apiary.—191. Out-door Apiary.—192. Bee-house.—193. Cabinet bee-houses.—194. Form and material of hives.—195. Village hive.—196. English hive.—197. Various forms of hives.—198. Various forms of bee-boxes.—199. Bee-dress and other accessories of apiculture.—200. Purchase of hives.—201. Honey harvest.—202. Honey and wax important articles of commerce.—203. Various sorts of wild honey.—204. Periodical migration of bees.—205. Poisoned honey.—206. Maladies of bees.—207. Curious case of abortive brood.—208. Superstition of bee cultivators.—209. Enemies of bees.—210. Attacks of bees when provoked.—211. Anecdote of Mungo Park.—212. Anecdote of Thorley.—213. Bee wars.—214. Curious case of a battle.

188. APICULTURE is the name given to the art by which the products of the industry of the bee are augmented in quantity, improved in quality, and rendered subservient to the uses of man.

189. The most favourable localities for the practice of apiculture are of course those of which the climate is suitable to the habits and character of the insect, and which most abound in those vegetable productions on which it loves to feed. Among these the principal are saintfoin, Dutch clover (*trifolium repens*), buckwheat, rape, honeysuckle, clover (*trifolium pratense*), and yellow trefoil (*medicago lupulina*). According to Dr. Bevan, the earliest

resources of the bee are, however, the *willow, hazel, osier, poplar,
sycamore,* and *plane;* to which may be added, the *snow-drop,
crocus, white alyssum, laurustinus, orange* and *lemon trees,
gooseberry* and *currant* and *raspberry bushes, sweet marjoram,
winter-savory, thyme,* and *mint.* In a word, fruit-trees and green-
house plants and shrubs in general, such especially as abound in
ornamental grounds, all constitute a part of bee-pasturage.

> " First the gray willows' glossy pearls they steal,
> Or rob the hazel of its golden meal ;
> While the gay crocus and the violet blue
> Yield to the flexile trunk ambrosial dew.
>
> EVANS, *quoted by* BEVAN.

An undulating country is highly favourable to the bee.

190. The apiary should be near the dwelling-house, in the garden,
and in a position sheltered from unfavourable winds. The farm
and poultry-yard should be avoided, as well as too great proximity
to railways, forges, factories, bakehouses, workshops, and the like.
The bee loves tranquil spots, planted with ornamental shrubs
and fruit-trees, and sown with sweet flowers, such as mignonette,
thyme, mint, rosemary, &c. The aspect of the apiary may be
east, west, or south, according as the one or other affords best
shelter, but never north.

191. The hives should be placed on separate stands, a few feet
apart, should be clear of any wall or fence, and elevated eighteen
inches or two feet above the ground.

Hives are sometimes assembled together in the open air,
forming an out-door apiary, such as is shown in fig. 54, p.1, in
which case they are generally made of straw, and protected in cold
weather by straw roofs, but sometimes also formed of wooden
boxes, as shown in the figure.

This arrangement, having the advantage of simplicity and
cheapness, is most commonly adopted, especially by those to whom
economy is important, and in warm climates where shelter is less
necessary.

192. Under other circumstances bee-houses are much more
strongly recommended, as well for comfort and convenience as for
security. The bee-house, one form of which is shown in fig. 55,
p. 33, consists of two or more rows of shelves, established one
above the other, on which the hives are placed at distances of
from twelve to eighteen inches apart, so that the bee-doors shall
be from two to three feet asunder. The house should be thatched
not only on the roof but on the sides and ends. A passage should
be provided for approaching the hives behind, and windows in the
side for ventilation.

193. A form called the Cabinet bee-house is shown in fig. 56,

p. 65, where B B are doors, one of which is glazed, and A a pipe of tin or caoutchouc, by which the bees have ingress and egress.

194. Hives have been constructed of different materials, as straw, osiers, rushes, sedges, wood, and earthenware; and of still more various forms, some being bell-shaped or conical, some cylindrical, some square in their section, some with rectangular and some with oblique tops, being internally divided by comb-frames fixed or movable, by shelves, and other expedients.

Their forms of structure depend in some degree upon the object of the proprietors. When apiculture is prosecuted on a large scale for the produce of honey and wax, as articles of trade, the foreign cultivators prefer hives of the most simple forms and most easy construction, and those from which the products can be obtained with most facility. The material preferred is, generally, straw or rushes. The process of making such a hive is indicated in fig. 57.

Fig. 57.—Process of making a straw hive.

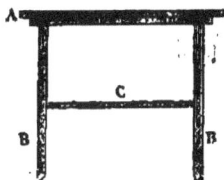

Fig. 61.—Movable comb-frame of the village hive.

Fig. 59.—Top of the cylindrical body of the village hive.

195. The bell-shaped straw hive, called the village hive, represented on the right of fig. 58, p. 49, is cylindrical in the body, and surmounted by a bell-shaped cap. The top of the cylindrical body is covered by a frame of bars, shown separately in fig. 59, and the cap itself is shown in fig. 60.

Fig. 60.—Cap of the village hive.

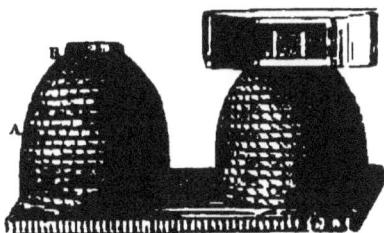

Fig. 62.—Dewhurst's hive.

One of the movable comb-frames is shown in fig. 61, where A is the vertical section of the stage, shown by plan in fig. 59; B the uprights, and c a shelf shown in vertical section.

196. The English hive of Dewhurst, having a box at the top, is shown in fig. 62; where A is the body of the hive, B the opening at the top, and C the box provided with shutters.

197. In fig. 63, p. 81, is shown a form of straw hive used in Scotland, and in fig. 64 the Radouan hive, similar in form to the village hive, but provided with movable pieces, by placing which successively below it, its elevation can be gradually augmented without disturbing the superior part, so as to give increased space to the bees and prevent the issue of swarms.

A form of hive much used in the South of France, and known to French apiarists as the Vulgar Hive (Ruche Vulgaire), is shown on the left of fig. 58, p. 49, in the process of transferring the bees from one hive to another.

A form of cork hive used in the South of France is shown in

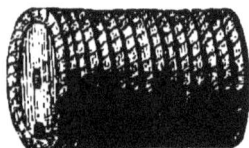

Fig. 66.—Cylindrical hive (Switzerland and Italy).

Fig. 67.—Della Rocca hive (Greece and Turkey).

fig. 65; and a cylindrical hive with its axis horizontal, much

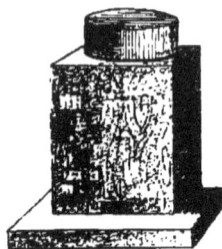

Fig. 68. — Murie's bee-box, with cylindrical cap (French).

Fig. 69.—De Frarière's garden hive.

used in Switzerland and Italy, is shown in fig. 66.

In Greece and Turkey a hive of earthenware, known as that of della Rocca, is much used, fig. 67.

Straw hives have the advantage over wooden boxes in being better non-conductors of heat, and therefore preventing immoderate cold in winter and immoderate heat in summer in the

Fig. 70.—Patteau's bee-box, with horizontal divisions.

Fig. 71.—Gélieu's bee-box, with vertical division.

interior. They are on this account preferred where the apiary is uncovered.

198. When apiculture is practised partly for the purpose of

Fig. 72.—Feburier's bee-box, with vertical division and sloping roof.

Fig. 73.—Huber's experimental leaf-hive.

observing the habits of the insect, boxes with divisions and movable comb-frames, with glazed openings and other like contrivances, are used. These *bee-boxes*, as they are called, are

Fig. 74.—Debeauvoy's bee-box, with sloping roof and shelves.

Fig. 75.—Vertical frame of box shown in fig. 74.

infinitely various in form, and although our limits will not allow us to enter into the details of the advantages derived from them

Fig. 77.—Debeauvoy's box, with vertical frames.

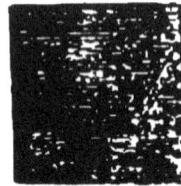

Fig. 78.—Shutter of box shown in fig. 77.

by their inventors and contrivers, it will nevertheless be useful to show the forms of those most generally used.

The common bee-box used in the South of France is shown in fig. 76, p. 17, the cover c being hinged, so as to be capable of being raised at pleasure. The process of transferring the bees

102

Fig. 78.*—Lefebvre's box, with movable frames. A, a frame drawn out.

Fig. 79.—Hamet's bee-box, with oblique horizontal divisions.

Fig. 80.—One of the divisions by which fig. 79 is elevated, with a movable frame, A, drawn out.

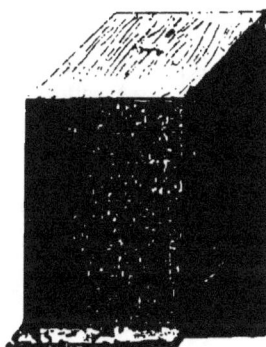

Fig. 81.—Hamet's bee-box, with divisions and movable frames.

from one hive to another by smoking them, is indicated, and also the method of hiving a swarm.

Fig. 82.—Uprights of fig. 81.

Fig. 83.—Frame of fig. 81.

Fig. 84.—Side of fig. 81, with its movable frame.

199. In the practical details of apiculture there are many

accessories, some of which are of occasional, and others of constant use.

The *bee-dress*, fig. 86, is a sort of armour, by which the operator is protected from all hostile attacks of the insect. It is usually made of Scotch gauze, or catgut, and so formed as to inclose the head, neck, and shoulders, as shown in fig. 76, p. 17, where a person invested with such a dress is represented in the act of hiving a swarm. It should have long sleeves to tie round the wrists over a pair of thick gloves, and the body should descend low enough to be tied round the waist. Thick woollen stockings and a woollen apron are recommended, the material being one from which the bee can readily withdraw its sting.

Fig. 87.   Fig. 88.

Knives of different forms (figs. 87, 88) should be provided, for the partial removal of the honey-combs, when the smothering process is not resorted to.

A bellows connected with a fumigator (fig. 89) for projecting tobacco-smoke into those parts of the combs from which it is desired to expel the bees, should be provided.

A hive with a handle for mixing swarms is often useful (fig. 90).

A basket, with an open bottom, placed over a tub for the purpose of draining the honey-combs, is also a convenient accessory (fig. 91).

200. A hive should, in general, be purchased in autumn, and its value will be pretty well ascertained by its weight. That of a good hive which will be sure to go through the winter, and to be productive in the ensuing season, should be from 25 to 30 lbs., and should contain about a peck of bees. If the weight be much greater than 30 lbs., a part of the honey may be advantageously taken out. Hives are to be preferred which are only a year old, and which have sent out no more than a single swarm. Such will be distinguished by the superior whiteness and purity of the combs. The transport should be made in cool weather, and should be conducted without shocks or jolts.

201. Honey should never be taken from any but the nearest and most populous hives. If they are provided with movable comb-frames, it is usual to make a partial harvest in May, the principal stores of the insect being collected between the middle of May and the end of June, the commencement and termination, however, varying three or four weeks, according to the climate peculiar to the locality.

Dr. Bevan recommends, as a general rule, that no honey should be taken from a colony the first year of its being planted.

Fig. 90.

Fig. 89.

Fig. 91.

To make a partial collection of honey, the hive is opened at the top or at the side, and the bees expelled from the combs by puffing tobacco-smoke upon them. The combs are then cut away with knives of suitable forms (figs. 87, 88). This operation requires to be performed with skill and care, so as to avoid as much as possible irritating the bees. To withdraw the queen from the part of the combs which are to be removed, the operator taps with his fingers on the opposite part of the hive, which will cause her majesty to run there, to ascertain the cause of the noise. If any bees are seen upon the combs removed, they may be brushed off with a feather, when they will generally return to the hive. The combs taken away are replaced either by empty ones, or by full combs taken from the lower part of the hive.

When hives are constructed on the principle of those shown in fig. 64, &c., consisting of several parts separable, laid one upon the other, the honey may be collected by causing the bees to desert the division intended to be removed by tapping on remote parts of the hive, and by projecting tobacco-smoke on them.

These operations may he performed in the day between ten and three o'clock. If the country be one rich in bee pasturage, a superior division of the hive may be taken away and replaced by an empty one, if the operation take place early in the season; and this latter may sometimes be again harvested before the close of the season, so as to obtain honey of the purest and finest quality.

But where the pasturage is not so rich, or where the operation is performed later in the season, it will be necessary either not to replace the division harvested, or to put the empty division at the bottom of the hive.

To collect the honey in the hives of the form represented in fig. 58, p. 49, called the vulgar hive, it is necessary either to expel the bees or to smother them.

To expel and transfer them to another hive, that which is to be harvested is inverted, as shown in fig. 58, p. 49, and over it is placed the hive to which the bees are to be transferred. The bees may be driven from one to the other, either by being smoked, as shown in fig. 76, p. 17, or by tapping upon the superior hive, fig. 58, p. 49.

If some bees remain in the hive to be harvested, they will voluntarily pass into the new hive by the arrangement represented in fig. 76, p. 17.

When the hive is harvested, either wholly or partially, by affecting the bees with temporary asphyxia, the process is as follows : after having beaten the black powder from a puff-ball of Lycoperdon, it is placed with some red charcoal in the fumigator, fig. 89, the nozzle of which is inserted at the door of the hive. The bellows being worked for five or six minutes, the bees will fall insensible from the hive, when the combs may be removed, wholly or partially, as the case may be. In twenty or thirty minutes the bees will revive, and re-enter the hive, or may be received in a new one if desired.

If it be not desired to preserve the bees, the hive may be placed over a pit into which they will fall, and where they may be buried.

To obtain honey of the first quality, the purest combs, containing neither bee-bread nor brood, being selected, are drained through a hair-sieve or osier-basket. Their product, called virgin honey, is limpid. It hardens and keeps if potted and put in a cool and dry place. Honey of inferior quality is obtained by pressing the residue of the combs, and exposing them to heat.

Whenever honey is collected, wax may also be obtained, but the latter substance may be separately collected at the close of the winter, by paring away the lower ranges of comb, taking away by the knife those which are old, black, and mouldy, and those which have been attacked by the moth. The wax is dissolved with boiling water, after which it is purified and collected in moulds of glazed pottery.

202. Honey and wax, the products of bee industry, form important articles of commerce in various parts of the world.

Although the production of wax is not confined to the bee,

nearly all of that article employed in Europe is of bee manufacture.

Although honey has lost much of its importance as an article of food, since the discovery and improvement of the fabrication of sugar, it is still regarded as a luxury, and of considerable value in this country, as the material out of which a wholesome vinous beverage is produced. In many inland parts of the continent where sugar is costly, few articles of rural economy could be less spared. In the Ukraine some of the peasants possess from 400 to 500 hives, and are said to make more profit of their bees than even of their corn. In Spain the nurture of bees is carried to a still greater extent; according to Mills, a single parish priest was known to possess the almost incredible number of 5000 hives.

The common hive-bee is the same, according to Latreille, in every part of Europe, except in some districts of Italy, where a species called the Apis ligustica of Spinola is kept. This species is also said to be cultivated in the Morea and the Ionian Isles. Honey, however, is also obtained from many other species of bees, as well wild as domesticated.

203. The rock honey of some parts of America, which is very thin and as clear as water, is the produce of wild bees, which suspend their clusters of thirty or forty waxen cells, resembling a bunch of grapes, from a rock. In South America large quantities of honey are collected from nests built in trees by the Trigona Amalthea and other species of this genus, under which, according to Kirby, should be included the Bamburos, to gather the honey of which the whole population in Ceylon make excursions into the woods.

According to Agara, one of the chief articles of food of the Paraguay Indians is wild honey.

Captain Green observes, that in the Island of Bourbon, where he was stationed for some time, there is a bee which produces honey much esteemed there, of a green colour, having the consistency of oil, and which, besides the usual sweetness of honey, has a remarkable fragrance. This green honey is exported to India in considerable quantities, where it bears a high price.

A species of bee called the Apis fasciata was probably cultivated ages before the present hive-bee was attended to. This species is still so extensively cultivated in Egypt that Niebuhr met on the hill between Cairo and Damietta a convoy of 4000 hives, which the apiarists of that country were transporting from a region where the season had passed, to one where the spring was later.

204. This periodical migration of bees is by no means of modern date. According to Columella, the Greeks used to send their

bee-hives at certain seasons of the year from Achaia into Attica, and a similar custom still prevails in Italy, and even in this country in the neighbourhood of heaths.

Among the domesticated species of bees may be also mentioned the Apis unicolor in Madagascar, the Apis Indica at Pondicherry and in Bengal, and the Apis Adansonii at Senegal.

Fabricius affirmed that the Apis Acraensis laboriosa, and others in the East and West Indies, might be domesticated with greater advantage than even the common hive-bee of Europe, called the Apis mellifica.

205. Honey is one of the class of aliments which requires to be used with some precaution, since not only are certain constitutions of body affected injuriously by it, even in its most natural and wholesome state, but it happens occasionally that the insects which collect it resort to poisonous flowers, which impart their noxious properties to the honey extracted from them.

Kirby mentions the case of a lady of his acquaintance upon whom ordinary honey acted like poison, and says, that he heard of instances in which death ensued from eating it.

But where the bee unfortunately resorts to poisonous plants, the consequences are not thus limited to individuals of peculiar idiosyncrasies. Dr. Barton has given a remarkable example of this. [*]

In the autumn and winter of the year 1790, an extensive mortality was produced amongst those who had partaken of the honey, collected in the neighbourhood of Philadelphia. The attention of the American government was excited by the general distress; a minute enquiry into the cause of the mortality ensued, and it was satisfactorily ascertained that the honey had been chiefly extracted from the flowers of *Kalmia latifolia*. Though the honey mentioned in Xenophon's well-known account of the effect of a particular sort, eaten by the Grecian soldiers during the celebrated retreat, after the death of the younger Cyrus, did not operate fatally, it gave those of the soldiers who ate it in small quantities the appearance of being intoxicated, and such as partook of it freely, of being mad or about to die, numbers lying on the ground as if after a defeat. A specimen of this honey, which still retains its deleterious properties, was sent to the Zoological Society in 1834 from Trebizond, on the Black Sea, by Keith E. Abbott, Esq.

206. The maladies of the bee proceed from three causes,—hunger, damp, and infection; all of which admit of prevention when the insect is maintained artificially.

---

[*] American Philosophical Transactions, vol. v. of the year 1790.

Dysentery is the malady which is at once the most dreaded by bee-owner, and the most easy to be prevented. It is always due to damp or to bad diet, such as impure honey and indigestible syrups. The remedies are consequently to place the hives in a dry situation, and to supply the insects with wholesome food, such as good honey mixed with a little generous wine. The greatest care should also be taken to remove such combs as may be rendered foul by excrement, and to clean the shelves in the bee-houses.

Among other maladies may be mentioned, diseases of the antennæ, vertigo, and abortive broods of eggs. These are generally produced by bad food, damp, and drafts of cold air. On that account some bee-cultivators reject the forms of hive or bee-houses having two doors on opposite sides, thus placed for the purpose of ventilation. This arrangement is never seen in the natural habitations of the insect.

207. Dr. Bevan mentions a case of abortive brood which occurred in one of Mr. Dunbar's hives. The colony had been very strong in the previous autumn, and possessed a fertile queen, but in the spring it failed, and did not swarm. On examination, he found the four central leaves of the hive (which was one of Huber's, fig. 73), full of abortive brood, by the presence of which the queen seemed to be paralysed, though she still laid a few eggs at the edge of the combs. As the population seemed gradually diminishing, Mr. Dunbar cut out the whole of the abortive brood, removed the old queen, and added an after swarm to the family. The conjoined bees soon betook themselves to work, replaced the old combs by new ones, and laid in an ample store of honey. This is an operation called *castration* by French apiculturists; and in all such cases it is prudent, in order to prevent contagion, to have the infected combs burnt or buried.

208. Butler, in his "Female Monarchy," relates a story of a credulous lady who devoted herself to the cultivation of bees. This person having gone to receive the sacrament, retained the consecrated wafer; and at the suggestion of a friend, more simple than herself, placed it in one of her diseased hives. The bee plague, according to her report, immediately ceased; honey accumulated; and, on examining the inside of the hive, she found there, to her astonishment and admiration, a waxen chapel, of wondrous architecture, supplied with an altar, and even with a steeple, and a set of bells, all constructed of the same material.

209. The most dangerous enemies of the bees are the larvæ of certain moths, which when once they take possession of a hive cannot be extirpated, and no remedy remains but to transport the entire population of the insect colony to a new habitation.

The bee-louse, an insect about the size of a flea, often infests populous hives, so as greatly to annoy the bees by fixing itself upon them. Sometimes two or more attach themselves to a single bee, making it restless and indisposed for its usual industry.

A magnified view of one of these parasites is shown in fig. 92, as seen from above; and in fig. 93, as seen from below.

Fig. 92.—Bee Louse, seen from above.

Fig. 93.—Bee-Louse, seen from below.

That universal plunderer the wasp, and his formidable congener the hornet, often seize and devour them; sometimes ripping open their body to come at the honey, and at others carrying off that part in which it is situated. Wasps frequently take possession of a hive, having either destroyed or driven away its inhabitants, and consume all the honey it contains. Nay, there are certain idlers of their own species, called by apiarists, corsair bees, which plunder the hives of the industrious.

210. Examples have been already cited, in which bees have manifested peculiar personal antipathies, which have been ascribed, in the cases mentioned, to some odour, offensive to the insect, proceeding from the obnoxious individuals. Independently, however, of such general causes of hostility, the insects are sometimes provoked against even their best friends and most familiar acquaintances, by occasional circumstances. Kirby relates, that although he was generally exempt from their hostility, he could not venture with impunity to put them out of humour. Thus happening one day, during the season when asparagus was in blossom, to pass among the beds, which were crowded with bees, he discomposed them so much that he was obliged to make a hasty retreat, pursued by a swarm of his offended friends.

211. In Mungo Park's last mission to Africa, he was much annoyed by bees. His people, searching for honey, having disturbed a large colony of them, the insects sallied forth by myriads, and attacking men and beasts indiscriminately, put them all to the rout. One horse and six asses were killed or missing in consequence of their attack, and for half an hour the bees seem to have completely put an end to their journey. Isaacs, upon

another occasion, lost one of his asses, and one of his men was almost killed by them.*

212. Bees, however, as we have already observed, are not usually ill-tempered ; and, if not molested, are generally inoffensive. Thorley relates,† that a maid servant, who assisted him in hiving a swarm, being rather afraid, put a linen cloth as a defence over her head and shoulders. When the bees were shaken from the tree on which they had alighted, the queen probably settled upon this cloth, for the whole swarm covered it, and then getting under it, spread themselves over her face, neck, and bosom, so that when the cloth was removed, she was quite a spectacle. She was with great difficulty kept from running off with all the bees upon her. But at length her master quieted her fears, and began to search for the queen. He succeeded, and expected that when he put her into the hive the bees would follow. He was, however, in the first instance disappointed, for they did not stir. Upon examining the cluster again, he found a second queen, or probably the former one, which had flown back to the swarm. Having seized her, he placed her in the hive, and kept her there. The bees soon missed her, and crowded into the hive after her, so that, in two or three minutes, not one remained on the poor frightened girl. After this escape she became quite a heroine, and would undertake the most hazardous employment about the hives.

213. The duels of rival queens have been already mentioned. Similar combats take place occasionally between the workers of one hive and those of another. Nor are such wars confined to single combats. General actions take place now and then between neighbouring colonies. This occurs when one takes a fancy to a hive which another has pre-occupied. Reaumur witnessed one of these battles, which lasted a whole afternoon, and in which great numbers fell on the one side and the other. In such cases, each combatant selects his opponent, and the victorious one flies away with the slain body of its enemy between its legs. After making a short flight thus, she deposits it on the ground, and rests near it, standing on her four anterior legs, and rubbing the two hinder legs against each other, as though she enjoyed the sight of her victim.

214. The following account of a bee battle was published in a Carlisle newspaper. A swarm of bees flying over a garden, where a newly tenanted hive was placed, suddenly stopped in their flight, and, descending, settled upon the hive, completely covering it. In a little time, they began to make their way to the door, and poured into it in such numbers, that it became completely

* Park's Last Mission, 153, 297.   † Thorley, 150.

filled. A loud humming noise was heard, and the work of destruction immediately ensued. The winged combatants sallied forth from the hive until it became entirely emptied, and a ferocious battle commenced in the air between the besiegers and the besieged. These intrepid warriors were so numerous, that they literally darkened the sky overhead like a cloud. Meanwhile, the destructive battle raged with great fury on both sides, and the ground beneath was covered with the killed and wounded. Hundreds were seen dispersed on the ground, lying dead, or crawling about in a disabled state. To one party at length the palm of victory was awarded, and they settled upon a branch of an adjoining tree, from which they were removed to the deserted hive, of which they took quiet possession, and commenced and continued their usual industry.

Fig. 1.—The Termes Embia.

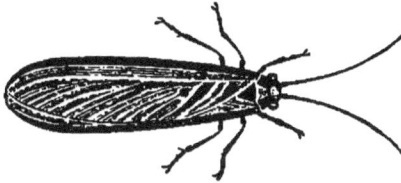

Fig. 2.—The Termes Fatalis, or Bellicosus, with wings folded.

Fig. 3.—Termes Fatalis, or Bellicosus, with wings expanded.   Fig. 4.—The King.

THE WHITE ANTS.

THEIR MANNERS AND HABITS.

CHAPTER I.

1. Their classification.—2. Their mischievous habits.—3. The constitution of their societies. — 4. Chiefly confined to the tropics.—5. Figures of the king and queen.—6. Of the workers and soldiers.—7. Treatment of the king and queen.—8. Habits of the workers.—9. Of the soldiers.—10. The nymphs.—11. Physiological characters. —12. First establishment of a colony.—13. Their use as food and medicine.—14. The election of the king and queen.—15. Their subsequent treatment.—16. The impregnation of the queen. —17. Figure of the pregnant queen.—18. Her vast fertility.—19. Care bestowed upon her eggs by the workers.—20. The royal body-guard.—21. The habitation of the colony.—22. Process of its construction.—23. Its chambers, corridors, and approaches.—24.

THE WHITE ANTS.

1. Of all the classes of insects which live in organised societies, the most remarkable after the bee are the family Termitinæ, popu- larly known under the name of white ants, though they have little in common with the ant, except their social character and habits.

Much discordance has prevailed among naturalists respecting their history and classification. They were assigned by Linnæus to the order Aptera, or wingless insects. More exact observation has, however, proved this to be erroneous; since, in the perfect state, they possess membranous wings like those of the dragon-fly, which being four in number, they have been more correctly assigned to the order Neuroptera. Kirby regards them as forming, together with the ants, a link between the orders Neuroptera and Hymenoptera, being allied to the latter by their social instincts.

2. Scarcely less remarkable than the bee in their social organisa- tion, they differ from that insect inasmuch as while the labours of the latter are attended with no evil to mankind, but are, on the contrary, productive of an eminently useful and agreeable article of food, the Termites, so far as naturalists have yet dis- covered, are productive of nothing but extensive and unmitigated mischief.

3. These insects live in societies, each of which consists of countless numbers of individuals, the large majority of which are apterous, or wingless. Two individuals only in each society, a male and a female, or according to some, a king and a queen, are winged, and these alone in the entire society are specimens of the perfect insect. The general form of their bodies is shown in

98

fig. 1 and fig. 2; the former representing the species called the *Termes embia*, with its wings expanded, and the latter the *Termes fatalis* or *bellicosus*, with its wings folded.

4. With the exception of two or three small species, such as the *Termes lucifugus*, described by Latreille and Rossi; the *Termes flavicollis*, described by Fabricius; and the *Termes flavipes*, described by Kollar, these insects are confined chiefly to the tropics.

5. Each society consists of five orders of individuals—

       I. The queen or female.
      II. The king or male.
     III. The workers.
     IV. The nymphs.
      V. The neuters or soldiers.

The *Termes bellicosus* or *fatalis*, which is represented in fig. 2, with wings folded, is shown in fig. 3 with wings expanded.

The king or male, which never changes its form after losing its wings, is represented in fig. 4.

6. The worker is represented in its natural size in fig. 5, and the soldier in fig. 6.

A magnified view of the worker is given in fig. 7, and a similar magnified view of the foreeps of the soldier in fig. 8.

7. The king and queen are privileged individuals, surrounded with all the respect and consideration, and receiving all the attendance and honours, due to sovereigns. Exempted from all participation in the common industry of the society, they are wholly devoted to increase and multiplication, the queen being endowed with the most unbounded fertility. Though upon first passing from the pupa state they have four wings, they lose these appendages almost immediately, and during the period of their sovereignty they are wingless. They are distinguished from the inferior members of the society by the possession of organs of vision, in the form of large and prominent eyes, their subjects being all of them blind.

8. The workers are by far the most numerous members of the society, being about a hundred times greater in number than the soldiers. Their bodies also, fig. 5, are less than those of the soldiers, the latter being less than those of the sovereigns. The entire industrial business of the society is performed by the workers. They erect the common habitation, and keep it in repair. They forage and collect provisions for the society. They attend upon the sovereigns, and carry away the eggs of the queen, as fast as she deposits them, to chambers which they previously prepare for them. They maintain these chambers in order, and when the eggs are hatched, they perform the part of nurses to the young,

feeding and tending them until they have attained sufficient growth to provide for themselves.

9. The soldiers, of whom, as already observed, there is not more than one to every hundred workers, are distinguished by their long and large heads, armed with long pointed mandibles. Their duty, as their title implies, is confined to the defence of the society and of their common habitation, when attacked by enemies.

10. The nymphs differ so little from the workers, that they would be confounded with them, but that they have the rudiments of wings, or, more strictly speaking, wings already formed, folded up in wing cases. These escaped the notice of the earliest observers, having been distinguished by Latreille.

11. Naturalists are not agreed as to the physiological character of these three classes of the society. Some consider the workers as the larvæ which, at a certain advanced period of their growth, are metamorphosed into the nymphs, which themselves finally pass into the state of the perfect winged insect.

According to Kirby, the soldiers correspond to the neuters in other societies of insects. As he observes, however, they differ from the neuters of the societies of Hymenoptera, which are a sort of sterile females. He conjectures that the soldiers may be the larvæ which are finally transformed into the perfect male insect. Great differences of opinion, however, prevail on this subject among entomologists.

For our present purpose, these doubtful questions, whatever interest they may have for naturalists, are altogether unimportant. What we desire at present to direct attention to, is the curious manners and habits of these insects, which have been ascertained by many eminent naturalists, and have been described with great minuteness by Smeathman in the seventy-first volume of the Philosophical Transactions, from whose memoir we shall here borrow largely.

According to Smeathman, the following is the manner in which the establishment of each colony takes place.

12. The pupæ or nymphs, which compose, as has been stated, part of a society, are transformed into the perfect insect, their wings being fully developed and liberated from the wing cases soon after the first tornado, which takes place at the close of the dry season, and harbingers the periodical rains. The insects, thus perfected, issue forth from their habitation in the evening, in numbers literally countless, swarming after the manner of bees. Borne upon their ample wings, and transported by the wind, they fill the air, entering houses, extinguishing lights, and being sometimes driven on board ships which happen to be near the shore. The next morning they are seen covering the surface of the earth

and waters, deprived of the wings which enabled them, for a moment, to escape their numerous enemies. They are now seen as large maggots, and, from being the most active, industrious, and sagacious of creatures, are become utterly helpless and cowardly, and fall a prey to innumerable enemies, to the smallest of which they do not attempt to offer the least resistance. Various insects, and especially ants, lie in wait for them; beasts, birds, and reptiles, and even man himself, all feed upon them, so that not one pair in many millions make their escape in safety, and fulfil the first law of nature by becoming the parents of a new community. At this time they may be seen running upon the ground, the male pursuing the female, and sometimes two pursuing one, and contending with the greatest eagerness for the prize, their passion rendering them regardless of the many dangers with which they are surrounded.

13. Mr. König, in an essay upon these insects, read before the society of naturalists at Berlin, says that, in some parts of the East Indies, the queens are given alive to old men for strengthening the back, and that the natives have a method of catching the winged insects, which he calls females, before the time of emigration. They make two holes in the nest; the one to windward and the other to leeward. At the leeward opening, they place the mouth of a pot, previously rubbed with an aromatic herb, called Bergera, which is more valued there than the laurel in Europe. On the windward side they light a fire of stinking materials, the smoke of which not only drives these insects into the pots, but frequently the hooded snakes also, on which account they are obliged to be cautious in removing them. By this method they catch great quantities, of which they make with flour a variety of pastry, which they can afford to sell very cheap to the poorer ranks of people. Mr. König adds, that in seasons when this kind of food is very plentiful, the too great use of it brings on an epidemic cholic and dysentery, which kills in two or three hours.

Mr. Smeathman says, that he did not find the Africans so ingenious in procuring or dressing them. They are content with a very small part of those which, at the time of swarming, or rather of emigration, fall into the neighbouring waters, which they skim off with calabashes, bringing large kettles full of them to their habitations, and parch them in iron pots over a gentle fire, stirring them about as is usually done in roasting coffee. In that state, without sauce or any other addition, they serve them as delicious food, and put them by handfuls into their mouths, as we do comfits. Smeathman ate them dressed in this way several times, and thought them delicate, nourishing, and wholesome. They are something sweeter, but not so fat or cloying, as the caterpillar or maggot of the palm-tree snout beetle, which is served up at all the luxurious tables of West Indian epicures, particularly of the French, as the greatest dainty of the Western World.

14. Troops of workers, apparently deprived of their king and queen, which are constantly prowling about, occasionally encounter one of these pairs, to which they offer their homage, and seem to elect them as the sovereigns of their community, or the parents of the colony which they are about to establish. All the individuals of such a swarm, who are not so fortunate as to become the objects of such an election, eventually perish under the attacks of the enemies above mentioned, and probably never survive the day which follows the evening of their swarming.

15. So soon as this election has been made, the workers begin to enclose their new rulers in a small chamber of clay, suited to their size, the entrances to which are only large enough to admit themselves and the soldiers, but much too small for the royal pair to pass through, so that their state of royalty is a state of confinement, and so continues during the remainder of their lives.

16. The impregnation of the female is supposed to take place after this confinement, and she soon begins to furnish the infant colony with new inhabitants. The care of feeding her and her male companion devolves upon the workers, who supply them both with every thing that they want. As she increases in dimensions, they keep enlarging the cell in which she is detained. When the business of oviposition commences, they take the eggs from the female, and deposit them in the nurseries. Her abdomen now begins gradually to extend, till, in process of time, it is enlarged to 1500 or 2000 times the size of the rest of her body, and her bulk equals that of 20000 or 30000 workers.

17. A drawing of the pregnant queen in her natural size is given in fig. 9.

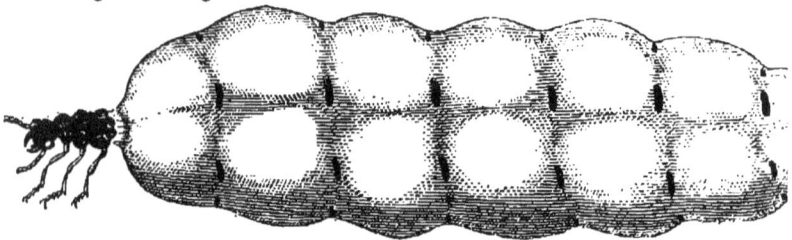

Fig. 9.—The Pregnant Queen.

18. The abdomen, often more than three inches in length, is now a vast matrix of eggs, which make long circumvolutions through numberless slender serpentine vessels : it is also remark-

able for its peristaltic motion (in this resembling the female ant), which, like the undulations of water, produces a perpetual and successive rise and fall over its whole surface, and occasions a constant extrusion of the eggs, amounting sometimes in old females to sixty in a minute, or eighty thousand and upwards in twenty-four hours. As these females live two years in their perfect state, how astonishing must be the number produced in that time!

19. This incessant extrusion of eggs must call for the attention of a large number of the workers in the royal chamber (and indeed it is always full of them), to take them as they come forth and carry them to the nurseries; in which, when hatched, they are provided with food, and receive every necessary attention until they are able to shift for themselves. One remarkable circumstance attends these nurseries. They are always covered with a kind of mould, amongst which arise numerous globules about the size of a small pin's head. This probably is a species of Mucor; and by Mr. König, who found them also in nests of an East India species of Termes, is conjectured to be the food of the larvæ.

20. The royal cell has in it a kind of body-guard to the royal pair that inhabit it; and the surrounding apartments always contain many, both labourers and soldiers in waiting, that they may successively attend upon and defend the common father and mother on whose safety depend the happiness and even existence of the whole community, and whom these faithful subjects never abandon, even in their last distress.

21. The habitations of the Termites, which are generally of considerable magnitude, vary in form, arrangement, and position, according to the species. Those of the *Termes bellicosus*, described above, have generally a sugar-loaf or hay-cock form, and are from ten to twelve feet high. In the parts of Africa where the insect prevails, these structures are so numerous that it is scarcely possible to find a spot from which they are not visible in all directions within fifty or sixty yards. In the neighbourhood of Senegal, according to Adanson, their number and magnitude is so great that they cannot be distinguished from the native villages.

22. When first erected, the external surfaces of these conical-shaped habitations consist of naked clay, but in these fertile climates the seeds of herbage transported by the wind are speedily deposited upon them, which germinating soon clothe them with the same vegetation as that which covers the surrounding soil, and when in the dry and warm season this vegetable covering is scorched, they assume the appearance of large hay-cocks.

23. These vast mounds are formed of earth which has been excavated by the workers from extensive tunnels which have

been carried beneath the ground surrounding their base, and which supply covered ways by which the workers are enabled to go forth in quest of provisions. The interior of the mounds themselves are of most curious and complicated structure, consisting of a variety of chambers and corridors, formed with the most consummate art, and adapted in shape and size to the respective purposes to which they are assigned in the general economy of the colony.

24. In the superior part of the mound, a dome is constructed, surmounting the habitations of the animals so as effectively to shelter them from the vicissitudes of weather. This may be seen in the vertical section of one of these mounds, shown in fig. 10. The exterior covering of this dome is much stronger than the internal structure beneath it, which constitutes the habitation of the colony, and which is divided with surprising regularity and contrivance into a vast number of chambers, one of which is appropriated to the sovereigns, and the others distributed among the soldiers, the workers, as nurseries, and as store-rooms.

The process by which these conical structures are raised is thus described.

25. The habitation makes its first appearance as one or two small sugar-loaf-shaped mounds about a foot in height. While these are gradually increasing in height and magnitude, others begin to appear near them, which likewise increase in number; and by the enlargement of their basis, they at length coalesce at the lower parts. The middle mounds are always the highest, and the largest, and by gradually filling up the intermediate space by the enlargement of the bases of the several mounds, a single mound, with various sugar-loaf-shaped masses of less magnitudes growing out of it, is produced, as shown in fig. 10.

a a a. Turrets by which their hills are raised and enlarged.

2. A section of 1, as it would appear on being cut down through the middle, from the top to the bottom, a foot lower than the surface of the ground.

A A. An horizontal line from A on the left, and a perpendicular line from A at the bottom will intersect each other at the royal chamber.

The darker shades near it are the empty apartments and passages, which, it seems, are left so for the attendants on the king and queen, who, when old, may require near one hundred thousand to wait on them every day.

The parts which are least shaded and dotted, are the nurseries, surrounded, like the royal chamber, by empty passages on all sides, for the more easy access to them with

104

Fig. 10.

View of the Habitations of the White Ants, reproduced from the original
drawing of Smeathman, engraved in the " Phil. Trans.," vol. lxxi.

the eggs from the queen, the provision for the young, &c.
N.B. The magazines of provisions are situated without any
seeming order, among the vacant passages which surround the
nurseries.

B. The top of the interior building, which often seems, from the
arches carried upward, to be adorned on the sides with
pinnacles.

C. The floor of the area or nave.

D D D. The large galleries which ascend from under all the
buildings spirally to the top.

E E. The bridge.

3. The first appearance of a hill-nest by two turrets.

4. A tree with the nest of the *Termites arborum*, with their
covered way.

F F F F. Covered ways of the *Termites arborum*.

5. The nest of the *Termites arborum*.

6. A nest of the *Termites bellieosi*, with Europeans on it.

7. A bull standing sentinel upon one of these nests.

G G G. The African palm-trees from the nuts of which is made the
Oleum palmæ.

26. When by the accumulation of these turrets the dome has
been completed, in which process the turrets supply the place of
scaffolding, the workers excavate the interior of them, and make
use of the clay in building the partitions and walls of the apartments constructed in the base of the mound which constitutes
their proper habitation, and also for erecting fresh turrets surmounting the mound and increasing its height. In this manner
the same clay, which, as has been already explained, was excavated
from the underground ways issuing around the mound, is used
several times over, just as are the posts and boards of a mason's
scaffolding.

27. When these mounds have attained a little more than half
their height, their tops being then flat, the bulls which are the
leaders of the herds of wild cattle which prevail in the surrounding country, are accustomed to mount upon them so as to obtain
a view of the surrounding plain: thus placed they act as sentinels
for the general herd which feeds and ruminates around them,
giving them notice of the approach of any danger. This circumstance supplies an incidental proof of the strength of these
structures.

28. Smeathman states that when he was in that country, and
desired to obtain a view of the sea to ascertain the approach
of vessels, he was in the habit of mounting with three or four
of his assistants upon the summits of these conical mounds,

106

the elevation of which was sufficient to enable him to obtain a satisfactory view.

29. The superior shell or dome by which the mound is sur-mounted is not only of use to protect the interior buildings from external violence and from the tropical rains, but, from its non-conducting quality, to preserve that uniform temperature within, which is necessary for hatching the eggs and cherishing the young.

30. The royal chamber appropriated to the sovereigns engrosses much of the attention and skill of their industrious subjects. It is generally placed about the centre of the base of the mound, at the level of the surrounding ground, and has the shape of half an egg divided by a plane at right angles to its axis passing a little below its centre. Thus the shape of this chamber is that which architects call a surmounted dome. Its magnitude is pro-portioned to that of the king and queen to whom it is appropriated. In the infant state of the colony, before the queen is advanced in pregnancy, the diameter of this room does not exceed an inch, but as the royal lady increases in the manner already described, the workers continually enlarge the room, until at length it attains a diameter of eight or nine inches. Its floor is perfectly level, and formed of clay about an inch thick. The roof is formed of a solid well-turned oval arch increasing in thickness from a quarter of an inch at the sides where it rests upon the floor.

31. The doors are cut in the wall, and made of a magnitude suitable to the entrance and exit of the soldiers and workers who attend on the royal pair, but much too small for the passage of the royal personages themselves.

32. This large chamber is surrounded by numerous others of less dimensions, and various shapes, all of which have arched roofs, some circular, and some elliptical. These chambers com-municate with each other by doors and corridors. Those which are immediately contiguous to the royal chamber are appropriated to the soldiers, who are in immediate attendance on the sovereign, and to the workers, whose duty it is to supply and attend the royal table, and to carry away the eggs as fast as they are laid by the queen.

33. Around these antechambers is another suite of apart-ments, consisting of store-rooms for provisions, chambers for the reception of the eggs, and nurseries for the young. The store-rooms are constructed like other parts of the habitation, with walls and partitions of clay, and are always amply supplied with provi-sions, which, to the naked eye, seem to consist of the raspings of wood and plants, which the workers destroy. Upon submitting them to the microscope, however, they are found to consist prin-

cipally of vegetable gums and inspissated juices. These are thrown together in masses of different appearance, some resembling the sugar on preserved fruits, some transparent, and others opaque, as is commonly seen in all parcels of gum.

The nurseries, on the other hand, are constructed in a manner totally different from the other rooms.

34. The walls and partitions of these consist entirely of wooden materials, cemented together with gum. These nurseries, in which the eggs are hatched, and the young secured, are small irregularly shaped rooms, none of which exceed half an inch in width.

35. When the nest is in the infant state, the nurseries are close to the royal chamber; but as in process of time the queen enlarges, it is necessary to enlarge the chamber for her accommodation; and as she then lays a great number of eggs, and requires a greater number of attendants, so it is necessary also to enlarge and increase the number of the antechambers; for which purpose the small nurseries first built] are taken to pieces, rebuilt a little further off a size larger, and their number increased.

36. Thus they continually enlarge their apartments, pull down, repair, or rebuild, according to their wants, with a degree of sagacity, regularity, and foresight, not observed among any other kind of animals or insects.

37. There is one remarkable circumstance attending the nurseries which ought not to be omitted. They are always found slightly overgrown with mould, and plentifully sprinkled with white globules, about the size of the head of a small pin. These may be at first mistaken for eggs; but submitting them to the microscope, they appear to be a species of mushroom, similar to the common mushroom, of the sort usually pickled. They appear, when whole, white like snow a little thawed and afterwards frozen; and, when bruised, seem to be composed of an infinite number of pellucid particles, having a nearly oval form, and difficult to be separated. The mouldiness seems to be composed of the same kind of substance. The nurseries are enclosed in chambers of clay, like the store-rooms, but much larger. In the early state of the nest, they are not bigger than a hazel-nut, but in large hills are much more spacious.

38. These magazines and nurseries, separated by small empty chambers and galleries, which run round them, or communicate from one to the other, are continued on all sides to the outer wall of the building, and reach up within it to two-thirds or three-fourths of its height. They do not, however, fill up the whole of the lower part of the hill, but are confined to the sides, leaving an open area in the middle, under the dome, very much resem-

bling the nave of an old cathedral, having its roof supported by three or four very large Gothic arches, of which those in the middle of the area are sometimes two and three feet high; but as they recede on each side, rapidly diminish, like the arches of aisles in perspective. A flattish roof, without perforation, in order to keep out the wet, if the dome should chance to be injured, covers the top of the assemblage of chambers, nurseries, &c.; and the area, which is above the royal chambers, has a flattish floor, also water-proof, and so contrived as to let any rain that may chance to get in, run off into the subterraneous passages which run from the basement of the lower apartments through the hill in various directions; and one of astonishing magnitude, often having a bore greater than that of a large piece of ordnance. Smeathman measured the diameter of one of these passages, which was perfectly cylindrical, and found it to be thirteen inches.

39. These subterraneous passages, or galleries, are lined very thick with the same kind of clay of which the hill is composed, and ascend the inside of the outer shell in a spiral manner, and winding round the whole building, up to the top, intersect each other at different heights, opening either immediately into the dome in various places, and into the interior building, the new turrets, &c., or communicating thereto by other galleries of different bores or diameter, either circular or oval.

From every part of these large galleries are various small tunnels or galleries, leading to different parts of the building. Under ground there are many which lead downward, by sloping descents, three or four feet perpendicular, among the gravel, from whence the workers cull the finer parts, which, being worked up in their mouths to the consistence of mortar, become that solid clay of which their hills and all their buildings, except their nurseries, are composed.

40. Other galleries again ascend, and lead out horizontally on every side, and are carried under ground, near to the surface, to a vast distance : for if you destroy all the nests within one hundred yards of your house, the inhabitants of those which are left unmolested farther off will, nevertheless, carry on their subterraneous galleries, and invade the goods and merchandises contained in it by sap and mine, and do great mischief, if you are not very circumspect.

41. But to return to the cities from whence these extraordinary expeditions and operations originate, it seems there is a degree of necessity for the galleries under the hills being thus large, being the great thoroughfares for all the labourers and soldiers going forth or returning upon any business whatever, whether

fetching clay, wood, water, or provisions; and they are certainly
well calculated for the purposes to which they are applied, by the
spiral slope which is given them; for if they were perpendicular,
the labourers would not be able to carry on their building with so
much facility, since they cannot ascend a perpendicular without
great difficulty, and the soldiers can scarcely do it at all.
It is on this account that sometimes a road, like a ledge, is
made on the perpendicular side of part of the building within
their hill, which is flat on the upper surface, and half an inch
wide, and ascends gradually like a staircase, or like those
roads which are cut on the sides of hills and mountains, that
would otherwise be inaccessible; by which, and similar contrivances, they travel with great facility to every interior part.

42. This too is probably the cause of their building a kind of
bridge of one vast arch, which answers the purpose of a flight of
stairs from the floor of the area to some opening on the side
of one of the columns which support the great arches. Such
bridges shorten the distance considerably to those labourers who
have the eggs to carry from the royal chamber to some of the
upper nurseries, which in some hills would be four or five feet in
the straightest line, and much more if carried through all the
winding passages which lead through the inner chambers and
apartments.

Smeathman found one of these bridges half an inch broad, a
quarter of an inch thick, and ten inches long, making the side of
an elliptic arch of proportional size; so that it is wonderful it
did not fall over or break by its own weight before they got it
joined to the side of the column above. It was strengthened by a
small arch at the bottom, and had a hollow or groove all the length
of the upper surface, either made purposely for the inhabitants
to travel over with more safety, or else, which is not improbable,
worn so by frequent treading.

43. "Consider," observes Kirby, "what incredible labour and
diligence, accompanied by the most unremitting activity and the
most unwearied celerity of movement, must be necessary to enable
these creatures to accomplish, their size considered, these truly
gigantic works. That such diminutive insects, for they are
scarcely the fourth of an inch in length, however numerous,
should, in the space of three or four years, be able to erect a
building twelve feet high, and of a proportionable bulk, covered by
a vast dome, adorned without by numerous pinnacles and turrets,
and sheltering under its ample arch myriads of vaulted apartments of various dimensions, and constructed of different materials
—that they should moreover excavate, in different directions, and
at different depths, innumerable subterranean roads or tunnels,

some twelve or thirteen inches in diameter, or throw an arch of stone over other roads leading from the metropolis into the adjoining country to the distance of several hundred feet—that they should project and finish the, for them, vast interior stair-cases or bridges lately described—and, finally, that the millions necessary to execute such Herculean labours, perpetually passing to and fro, should never interrupt or interfere with each other, is a miracle of nature, or rather of the Author of nature, far exceeding the most boasted works and structures of man : for, did these creatures equal him in size, retaining their usual instincts and activity, their buildings would soar to the astonishing height of more than half a mile, and their tunnels would expand to a magnificent cylinder of more than three hundred feet in diameter ; before which the pyramids of Egypt and the aqueducts of Rome would lose all their celebrity, and dwindle into nothings.

"The most elevated of the pyramids of Egypt is not more than 600 feet high, which, setting the average height of man at only five feet, is not more than 120 times the height of the workmen employed. Whereas the nests of the Termites being at least twelve feet high, and the insects themselves not exceeding a quarter of an inch in stature, their edifice is upwards of 500 times the height of the builders ; which, supposing them of human dimensions, would be more than half a mile. The shaft of the Roman aqueducts was lofty enough to permit a man on horseback to travel in them." *

44. The bodies of the Termites are generally soft and covered with a thin and delicate skin, and being blind, they are no match on the open ground for the ants who are endowed with vision, and whose bodies are invested in a strong horny shell. Whenever the Termites are accidentally dislodged from their subterraneous roads or dwellings, the various species of ants instantly seize them and drag them away to their nests as food for their young.

45. The Termites are therefore very solicitous about preserving their tunnels and vaulted roads in good repair. If some of them be accidentally demolished for a few inches in length, it is wonderful how speedily they rebuild it. At first, in their hurry, they advance into the open part for an inch or two, but stop so suddenly that it is very apparent that they are surprised, for although some run straight on until they get under the arch beyond the damaged part, most of them run as fast back, and very few of them will venture through that part of the track which is left uncovered. In a few minutes, however, they will be seen rebuilding the arch, and even if three or four yards in length have been destroyed, they will reconstruct it in a single day. If this be again destroyed,

* Kirby, vol. i. p. 434.

they will be seen as numerous as ever passing both ways along it, and they will again in like manner reconstruct it. But if the same part be destroyed several times successively, they will give up the point and build a new covered way in another direction. Nevertheless, if the old one should lead to some favourite source of plunder, they will, after a few days' interval, still reconstruct it, apparently in the hope that the cause of destruction will not again occur, nor will they in that case wholly abandon the undertaking unless their habitation itself be destroyed.

Fig. 5.—Worker.

Fig. 6.—Soldier.

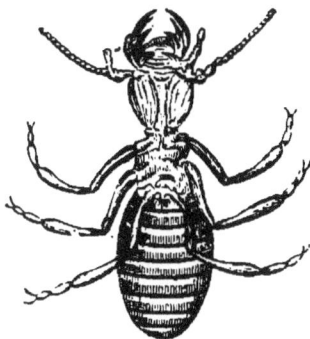

Fig. 7.—Worker, magnified.

Fig. 8.—Forceps of Soldier, magnified.

THE WHITE ANTS,

THEIR MANNERS AND HABITS.

CHAPTER II.

46. Turrets built by the Termes mordax and the Termes atrox.—47. Description of their structure.—48. Their king, queen, worker, and soldier.—49. Internal structure of their habitation.—50. Nests of the Termes arborum.—51. Process of their construction.—52. Hill nests on the Savannahs.—53. The Termes lucifugus—the organisation of their societies.—54. Habits of the workers and soldiers—the materials they use for building.—55. Their construction of tunnels.—56. Nests of the Termes arborum in the roofs of houses.—57. Destructive habits of the Termes hellicosus in excavating all species of wood-work—entire houses destroyed by them.—58. Curious process by which they fill with mortar the excavations which they make—destruction of Mr. Smeathman's microscope.—59. Destruction of shelves and wainscoting.—60. Their artful process to escape observation.—61. Anecdotes of them by Kœmpfer and Humboldt.—62. Destruction of the Governor's house at Calcutta—destruction by them of a British ship of the line.—63. Their manner of attacking timber in the open air—their wonderful power of destroying fallen timber.—64. The extraordinary behaviour of the soldiers when a nest is attacked.

46. A smaller species of Termites erect habitations, which, if they are of less dimensions, are not less curious in their structure. These buildings are upright cylinders, composed of a well-tempered black earth or clay, about three quarters of a yard high, and covered with a roof of the same material in the shape of a cone, whose base extends over and hangs down three or four inches wider than the perpendicular sides of the cylinder, so that most of them resemble in shape a round windmill, or still more closely the round towers which are so frequently seen in Ireland, and which have attracted so much attention on the part of antiquaries. Some of these roofs have so little elevation in the centre, that they have a close resemblance to certain species of mushroom.

After one of these turrets is finished, it is not altered or enlarged; but when no longer capable of containing the community, the foundation of another is laid within a few inches of it. Sometimes, though but rarely, the second is begun before the first is finished, and a third before they have completed the second: thus they will run up five or six of these turrets at the foot of a tree in the thick woods, and make a most singular group of buildings, as shown in fig. 11.

1 Nest of the *Termes mordax.*
2 Nest of the *Termes atrox.*
3 A turret with the roof begun.
4 A turret raised only about half its height.
5 A turret built upon one which has been thrown down.
6 6 A turret broken in two.

47. The turrets are so strongly built, that in case of violence they will much sooner overset from the foundations, and tear up the ground and solid earth, than break in the middle; and in that case the insects will frequently begin another turret and build it, as it were, through that which has fallen; for they will connect the cylinder below with the ground, and run up a new turret from its upper side, so that it will seem to rest upon the horizontal cylinder only.

Fig. 11.

The Turret Nests of the Termes Mordax and Termes Atrox.

48. In fig. 12 is represented the king or queen of the *Termes mordax*, in fig. 13 the worker, and in fig. 14 the soldier.

TERMES MORDAX.

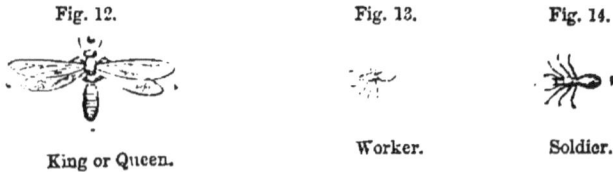

Fig. 12.	Fig. 13.	Fig. 14.

King or Queen. Worker. Soldier.

The building is divided into innumerable cells of irregular shapes; sometimes they are quadrangular or cubic, and sometimes pentagonal; but often the angles are so ill defined, that each half of a cell will be shaped like the inside of that shell which is called the sea-car.

49. Each cell has two or more entrances, but as there are no tunnels or galleries, no variety of apartments, no well-turned arches, wooden nurseries, &c., &c., as in the habitations already described, they are not calculated to excite the same degree of wonder, however admirable they may be considered without reference to other structures.

There are two sizes of these turret nests, built by two different species of Termites. The larger species, the *Termes atrox*, in its perfect state, measures one inch and three-tenths from the extremities of the wings on the one side to the extremities on the other. The lesser, *Termes mordax*, measures only eight-tenths of an inch from tip to tip.

50. The next kind of nests, built by another species of this genus, the *Termes arborum*, have very little resemblance to the former in shape or substance. These are generally spherical or oval, built in trees : sometimes they are established between, and sometimes surrounding, the branches, at the height of seventy or eighty feet; and are occasionally as large as a great sugar-cask.

51. They are composed of small particles of wood and the various gums and juices of trees, combined with, perhaps, those secreted by the animals themselves, worked by those little industrious creatures into a paste, and so moulded into innumerable little cells of different and irregular forms. These nests, with the immense quantity of inhabitants, young and old, with which they are at all times crowded, are used as food for young fowls, and especially for the rearing of Turkeys. These nests are very compact, and so strongly fixed to the boughs, that there is no detaching them but by cutting them in pieces, or sawing off the branch. They will even sustain the force of a tornado as long as the tree to

116

which they are attached. This species has the external habit, size, and almost the colour, of the *Termes atrox.*

52. There are some nests that resemble the hill-nests first described, built in those sandy plains called Savannahs. They are composed of black mud, raised from a few inches below the white sand, and are built in the form of an imperfect or bell-shaped cone, having their tops rounded. These are generally about four or five feet high. They seem to be inhabited by insects nearly as large as the *Termes bellicosus*, and differing very little from that species, except in colour, which is brighter.

53. The societies of *Termes lucifugus*, discovered by Latreille at Bourdeaux, are very numerous; but instead of making artificial nests, they make their lodgments in the trunks of pines and oaks, where the branches diverge from the tree. They eat the wood the nearest the bark without attacking the interior, and bore a vast number of holes and irregular galleries. That part of the wood appears moist, and is covered with little gelatinous particles, not unlike gum-arabic. These insects seem to be furnished with an acid of a very penetrating odour, which, perhaps, is useful to them in softening the wood. The soldiers in those societies are as about one to twenty-five of the labourers.

The anonymous author of the Observations on the Termites of Ceylon, seems to have discovered a sentry-box in his nests. " I found," says he, " in a very small cell in the middle of the solid mass, (a cell about half an inch in height, and very narrow,) a larva with an enormous head. Two of these individuals were in the same cell; one of the two seemed placed as sentinel at the entrance of the cell. I amused myself by forcing the door two or three times; the sentinel immediately appeared, and only retreated when the door was on the point to be stopped up, which was done in three minutes by the labourers."

54. Having thus given some idea of their habitations, we shall now direct our observations to the insects themselves, their manner of building, fighting, and marching, and to a more particular account of the vast mischief they cause to mankind.

It is a common character of the different species which have been noticed, that the workers and the soldiers never expose themselves in the open air, but invariably travel either under ground, or along the holes which they bore in trees and other substances. When in certain exceptional cases in quest of plunder they are compelled to move above ground, they make a vault with a coping of earth, or a tube, formed of that material with which they build their nests, along which they travel completely protected. The *Termes bellicosus* uses for this purpose the red, and the turret-builders black clay; whilst the *Termes arborum* employs for the purpose the ligneous substances of which their nests are composed.

55. With these materials they completely line most of the roads leading from their nests into the various parts of the country, and travel out and home with the utmost security in all kinds of weather. If they meet a rock or any other obstruction, they will make their way upon the surface, and for that purpose erect a covered way or arch, still of the same materials, continuing it with many windings and ramifications through large grooves, having, where it is possible, subterranean pipes running parallel with them, into which they sink, and save themselves, if their galleries above ground are destroyed by any violence, or the tread of men or animals alarms them. When any one chances by accident to enter any solitary grove, where the ground is pretty well covered with their arched galleries, they give the alarm by loud hissings, which he hears distinctly at every step he makes; soon after which he may examine their galleries in vain for the insects, which escape through little holes, just large enough for them, into their subterraneous roads. These galleries are large enough for them to pass and repass, so as to prevent any stoppages, and shelter them equally from light and air, as well as from their enemies, of which the ants, being the most numerous, are the most formidable.

56. The *Termites arborum*, those which build in trees, frequently establish their nests within the roofs and other parts of houses, to which they do considerable damage if not extirpated.

57. The larger species are, however, not only much more destructive, but more difficult to be guarded against, since they make their approaches chiefly under ground, descending below the foundations of houses and stores at several feet from the surface, and rising again either in the floors, or entering at the bottoms of the posts, of which the sides of the buildings are composed, bore quite through them, following the course of the fibres to the top, or making lateral perforations and cavities here and there as they proceed.

While some are employed in gutting the posts, others ascend from them, entering a rafter or some other part of the roof. If they once find the thatch, which seems to be a favourite food, they soon bring up wet clay, and build their pipes or galleries through the roof in various directions, as long as it will support them, sometimes eating the palm-tree leaves and branches of which it is composed, and perhaps (for variety seems very pleasing to them) the rattan or other running plant which is used as a cord to tie the various parts of the roof together, and to the posts which support it; thus, with the assistance of the rats, who,

118

during the rainy season, are apt to shelter themselves there, and to burrow through it, they very soon ruin the house by weakening the fastenings and exposing it to the wet. In the meantime, the posts will be perforated in every direction, as full of holes as that timber in the bottom of ships which has been bored by the worms; the fibrous and knotty parts, which are the hardest, being left to the last.

58. They sometimes, in carrying on this business, find that the post has some weight to support, and then, if it is a convenient track to the roof, or is itself a kind of wood agreeable to them, they bring their mortar, and fill all or most of the cavities, leaving the necessary roads through it, and as fast as they take away the wood, replace the vacancy with that material; which being worked together by them closer and more compactly than human strength or art could ram it, when the house is pulled to pieces, in order to examine if any of the posts are fit to be used again, those of the softer kinds are often found reduced almost to a shell, and all, or a greater part, transformed from wood to clay, as solid and as hard as many kinds of freestone used for building in England.

It is much the same when the *Termites bellicosi* get into a chest or trunk containing clothes and other things; if the weight above is great, or they are afraid of ants and other enemies, and have time, they carry their pipes through, and replace a great part with clay, running their galleries in various directions. The tree-Termites, indeed, when they get within a box, often make a nest there, and being once in possession destroy it at their leisure. They did so in a pyramidal box which contained the compound microscope of Mr. Smeathman. It was of mahogany, and he deposited it in the warehouse of Governor Campbell of Tobago, while he made a tour of a few months in the Leeward Islands. On his return he found that the Termites had done much mischief in the warehouse, and, among other things, had taken possession of the microscope, and eaten everything about it except the glass or metal, including the board on which the pedestal is fixed, with the drawers under it, and the things enclosed. The cells were built all round the pedestal and the tube, and attached to it on every side. All the glasses were covered with the wooden substance of their nests, and retained a cloud of a gummy nature upon them which was not easily got off, and the lacquer or burnish with which the brasswork was covered was totally spoiled.

Another party had taken a liking to a cask of Madeira, and had bored so as to discharge almost a pipe of fine old wine. If the large species of Africa (the *Termites bellicosi*) had been so long in the uninterrupted possession of such a warehouse, they would not have left twenty pounds weight of wood remaining of the whole building, and all that it contained.

59. These insects are not less expeditious in destroying the shelves, wainscotting, and other fixtures of a house, than the house itself. They are for ever piercing and boring in all directions, and sometimes go out of the broadside of one post into that of another joining to it; but they prefer, and always destroy the softer substances the first, and are particularly fond of pine and fir-boards, which they excavate and carry away with wonderful despatch and astonishing cunning; for, unless a shelf has something standing upon it, as a book, or anything else which may tempt them, they will not perforate the surface, but artfully preserve it quite whole, and eat away all the inside, except a few fibres which barely keep the two sides connected together, so that a piece of an inch board which appears solid to the eye will not weigh much more than two sheets of pasteboard of equal dimensions, after these animals have been a little while in possession of it.

60. In short the Termites are so insidious in their attacks, that we cannot be too much on our guard against them : they will sometimes begin and raise their works, especially in new houses, through the floor. If you destroy the work so begun, and make a fire upon the spot, the next night they will attempt to rise through another part; and, if they happen to emerge under a chest or trunk early in the night, will pierce the bottom, and destroy or spoil everything in it before morning. On these accounts care is taken by the inhabitants of the country to set all their chests and boxes upon stones or bricks, so as to leave the bottoms of such furniture some inches above the ground; which not only prevents these insects finding them out so readily, but preserves the bottoms from a corrosive damp which would strike from the earth through, and rot everything therein; a vast deal of vermin would also harbour under, such as cockroaches, centipedes, millepedes, scorpions, ants, and various other noisome insects.

61. Kœmpfer, speaking of the white ants of Japan, gives a remarkable instance of the rapidity with which these miners proceed. Upon rising one morning, he observed that one of their galleries, of the thickness of his little finger, had been formed across his table; and upon a further examination he found that they had bored a passage of that thickness up one foot of the table, formed a gallery across it, and then pierced down another foot into the floor; all this was done in the few hours that intervened between his retiring to rest and his rising. They make

their way also with the greatest ease into trunks and boxes, even though made of mahogany, and destroy papers and everything they contain, constructing their galleries and sometimes taking up their abode in them. Hence, as Humboldt informs us, throughout all the warmer parts of equinoctial America, where these and other destructive insects abound, it is infinitely rare to find papers which go fifty or sixty years back. In one night they will devour all the boots and shoes that are left in their way; cloth, linen, or books are equally to their taste; but they will not eat cotton. They entirely consumed a collection of insects made in India. In a word, scarcely anything but metal or stones comes amiss to them.

62. It is even asserted that the superb residence of the Governor-General at Calcutta, which cost the East India Company such immense sums, is now rapidly going to decay in consequence of the attacks of these insects. But not content with the dominions they have acquired, and the cities they have laid low on terra firma, encouraged by success, the white ants have also aimed at the sovereignty of the ocean, and once had the hardihood to attack even a British ship of the line ; and in spite of the efforts of the commander and his valiant crew, having boarded they got possession of her, and handled her so roughly, that when brought into port, being no longer fit for service, she was obliged to be broken up.

The ship here alluded to was the Albion, which was in such a condition from the attack of these insects, that had it not been firmly lashed together, it was thought she would have foundered on her voyage home. The late Mr. Kittoe stated that the *droguers* or *draguers*, a kind of lighter employed in the West Indies in collecting the sugar, sometimes so swarm with ants of the common kind, that they have no other way of getting rid of these troublesome insects than by sinking the vessel in shallow water.

63. When the Termites attack trees and branches in the open air, they sometimes vary their manner of doing it. If a stake in a hedge has not taken root and vegetated, it becomes their business to destroy it. If it has a good sound bark round it, they will enter at the bottom, and eat all but the bark, which will remain, and exhibit the appearance of a solid stick (which some vagrant colony of ants or other insects often shelter in, till the winds disperse it); but if they cannot trust the bark, they cover the whole stick with their mortar, and it then looks as if it had been dipped into thick mud that had been dried on. Under this covering they work, leaving no more of the stick and bark than is barely sufficient to support it, and frequently not the smallest particle, so that upon a very small tap with your walking stick,

121

the whole stake, though apparently as thick as your arm, and five or six feet long, loses its form, and, disappearing like a shadow, falls in small fragments at your feet. They generally enter the body of a large tree which has fallen through age, or been thrown down by violence, on the side next the ground, and eat away at their leisure within the bark, without giving themselves the trouble either to cover it on the outside, or to replace the wood which they have removed from within, being somehow sensible that there is no necessity for it. "Such excavated trees," says Mr. Smeathman, "deceived me two or three times in running; for, attempting to step two or three feet high, I might as well have attempted to step upon a cloud, and have come down with such unexpected violence, that, besides shaking my teeth and bones almost to dislocation, I have been precipitated head fore-most among the neighbouring trees and bushes." Sometimes, though seldom, the animals are known to attack living trees; but not before symptoms of mortification have appeared at the roots; since it is evident that these insects are intended in the order of nature to hasten the dissolution of such trees and vegetables as have arrived at their greatest maturity and perfection, and which would, by a tedious decay, serve only to encumber the face of the earth. This purpose they answer so effectually that nothing perishable escapes them, and it is almost impossible to leave any-thing penetrable upon the ground a long time in safety; for the odds are, put it where you will abroad, they will find it out before the following morning, and its destruction follows very soon of course. In consequence of this disposition, the woods never remain long encumbered with the fallen trunks of trees or their branches; and thus it is that the total destruction of deserted towns is so effectually completed, that in two or three years a thick wood fills the space; and, unless *iron-wood* posts have been made use of, not the least vestige of a house is to be discovered.

64. The first object of admiration, which strikes one upon opening their hills, is the behaviour of their soldiers. If you make a breach in a slight part of the building, and do it quickly, with a strong hoe or pick-axe, in the space of a few seconds a soldier will run out, and walk about the breach, as if to see whether the enemy is gone, or to examine what is the cause of the attack. He will sometimes go in again, as if to give the alarm; but most frequently, in a short time, is followed by two or three others, who run as fast as they can, straggling after one another, and are soon followed by a large body, who rush out as fast as the breach will permit them, and so they proceed, the number increasing, as long as any one continues battering their building. It is not easy to describe the rage and fury they show.

In their hurry they frequently miss their hold, and tumble down the sides of the hill, but recover themselves as quickly as possible; and being blind, bite everything they run against, and thus make a crackling noise, while some of them beat repeatedly with their forceps upon the building, and make a small vibrating noise, something shriller and quicker than the ticking of a watch. This noise can be distinguished at three or four feet distance, and continues for a minute at a time, with short intervals. While the attack proceeds, they are in the most violent bustle and agitation.

65. If they get hold of any one they will, in an instant, let out blood enough to weigh against their whole body; and if it is the leg they wound, you will see the stain upon the stocking extend an inch in width. They make their hooked jaws meet at the first stroke, and never quit their hold, but suffer themselves to be pulled away leg by leg, and piece after piece, without the least attempt to escape. On the other hand, keep out of their way, and give them no interruption, and they will, in less than half an hour, retire into the nest, as if they supposed the wonderful monster that damaged their castle to be gone beyond their reach.

66. Before they are all got in, you will see the labourers in motion, and hastening in various directions towards the breach; every one with a burthen of mortar in his mouth ready tempered. This they stick upon the breach as fast as they come up, and do it with so much dispatch and facility, that although there are thousands, and even millions of them, they never stop or embarrass one another; and you are most agreeably deceived when, after an apparent scene of hurry and confusion, a regular wall arises, gradually filling up the chasm. While they are thus employed, almost all the soldiers are retired quite out of sight, except here and there one, who saunters about among six hundred or a thousand of the labourers, but never touches the mortar either to lift or carry it; one, in particular, places himself close to the wall they are building.

67. This soldier will turn himself leisurely on all sides, and every now and then, at intervals of a minute or two, lift up his head, and with his forceps beat upon the building, and make the vibrating noise before mentioned; on which immediately a loud hiss, which appears to come from all the labourers, issues from within side the dome, and all the subterraneous caverns and passages: that it does come from the labourers is very evident, for you will see them all hasten at every such signal, redouble their pace, and work as fast again.

68. As the most interesting experiments become dull by repe-
123

tition or continuance, so the uniformity with which this business is carried on, though so very wonderful, at last satiates the mind. A renewal of the attack, however, instantly changes the scene, and gratifies our curiosity still more. At every stroke we hear a loud hiss ; and on the first the labourers run into the many pipes and galleries with which the building is perforated, which they do so quickly that they seem to vanish, for in a few seconds all are gone, and the soldiers rush out as numerous and as vindictive as before. On finding no enemy they return again leisurely into the hill, and very soon after the labourers appear loaded as at first, as active and as sedulous, with soldiers here and there among them, who act just in the same manner, one or other of them giving the signal to hasten the business. Thus the pleasure of seeing them come out to fight or to work alternately may be obtained as often as curiosity excites or time permits ; and it will certainly be found, that the one order never attempts to fight, or the other to work, let the emergency be ever so great.

69. We meet vast obstacles in examining the interior parts of these tumuli. In the first place the works, for instance, the apartments which surround the royal chamber and the nurseries, and indeed the whole internal fabric, are moist, and consequently the clay is very brittle ; they have also so close a connection, that they can only be seen as it were by piecemeal ; for having a kind of geometrical dependence or abutment against each other, the breaking of one arch pulls down two or three. To these obstacles must be added the obstinacy of the soldiers, who fight to the very last, disputing every inch of ground so well as often to drive away the negroes who are without shoes, and make white people bleed plentifully through their stockings. Neither can we let a building stand, so as to get a view of the interior parts without interruption, for while the soldiers are defending the outworks, the labourers keep barricading all the way against us, stopping up the different galleries and passages, which lead to the various apartments, particularly the royal chamber, all the entrances to which they fill up so artfully as not to let it be distinguishable, while it remains moist ; and externally it has no other appearance than that of a shapeless lump of clay. It is, however, easily found from its situation with respect to the other parts of the building, and by the crowds of labourers and soldiers which surround it, who show their loyalty and fidelity by dying under its walls. The royal chamber, in a large nest, is capacious enough to hold many hundreds of the attendants, besides the royal pair, and you always find it as full of them as it can hold. These faithful subjects never abandon their charge, even in the last distress, for whenever Mr. Smeathman took out the royal

ohamber from one of the hills, as he often did, and preserved it for some time in a large glass bowl, all the attendants continued running in one direction round the king and queen with the utmost solicitude, some of them stopping in every circuit at the head of the latter, as if to give her something; when they came to the extremity of the abdomen, they took the eggs from her, carrying them away, and piled them carefully together in some part of the chamber, or in the bowl under, or behind any pieces of broken clay, which lay most convenient for the purpose.

Some of these unhappy little creatures would ramble from the chamber as if to explore the cause of such a horrid ruin and catastrophe to their immense buildings, as it must appear to them; and after fruitless endeavours to get over the side of the bowl, return and mix with the crowd that continued running round their common parents to the last. Others, placing themselves along her side, would get hold of the queen's vast matrix with their jaws, and pull with all their strength, so as visibly to lift up the part which they fix at; but Mr. Smeathman who observed this, was unable to determine whether this pulling was with an intention to remove her body, or to stimulate her to move herself, or for any other purpose. After many ineffectual tugs, they would desist and join in the crowd running round, or assist some of those who are cutting off clay from the external parts of the chamber, or some of the fragments, and moistening it with the juices of their bodies, to begin to work a thin arched shell over the body of the queen, as if to exclude the air, or to hide her from the observation of some enemy. These, if not interrupted, before the next morning, completely cover her, leaving room enough within for great numbers to run about her.

The king, being very small in proportion to the queen, generally conceals himself under one side of her abdomen, except when he goes up to the queen's head, which he does now and then, but not so frequently as the rest.

70. If in your attack on the hill you stop short of the royal chamber, and cut down about half of the building, and leave open some thousands of galleries and chambers, they will all be shut up with thin sheets of clay before next morning. If even the whole is pulled down, and the different buildings are thrown in a confused heap of ruins, provided the king and queen are not destroyed or taken away, every interstice between the ruins, at which either cold or wet can possibly enter, will be so covered as to exclude both; and, if the animals are left undisturbed, in about a year they will raise the building to near its pristine size and grandeur.

71. The marching Termites are not less curious in their order

than those described before. This species seems much scarcer and larger than the *Termes bellicosus*. They are little known to the natives. Smeathman had an opportunity of observing them by mere accident; one day, having made an excursion with his gun up the river Camerankoes, on his return through the thick forest, while he was sauntering very silently in hopes of finding some sport, on a sudden he heard a loud hiss, which, on account of the many serpents in these countries, is a most alarming sound. The next step caused a repetition of the noise, which he soon recognised, and was rather surprised, seeing no covered ways or hills. The noise, however, led him a few paces from the path, where, to his great astonishment and pleasure, he saw an army of Termites coming out of a hole in the ground, which could not be above four or five inches wide. They came out in vast numbers, moving forward as fast seemingly as it was possible for them to march. In less than a yard from this place they divided into two streams or columns, composed chiefly of labourers, twelve or fifteen abreast, and crowded as close after one another as sheep in a drove, going straight forward, without deviating to the right or the left. Among these, here and there, one of the soldiers was to be seen, trudging along with them in the same manner, neither stopping nor turning; and as he carried his enormous large head with apparent difficulty, he appeared like a very large ox amongst a flock of sheep. While these were bustling along, a great many soldiers were to be seen spread about on both sides of the two lines of march, some a foot or two distant, standing still or sauntering about as if upon the look-out lest some enemy should suddenly come upon the workers. But the most extraordinary part of this march was the conduct of some others of the soldiers, who, having mounted the plants which grow thinly here and there in the thick shade, had placed themselves upon the points of the leaves, which were elevated ten or fifteen inches above the ground, and hung over the army marching below. Every now and then one or other of them beat with his forceps upon the leaf, and made the same sort of ticking noise, which he had so frequently observed to be made by the soldier who acts the part of surveyor or superintendent, when the labourers are at work repairing a breach made in one of the common hills of the *Termites bellicosi*. This signal among the marching white ants produced a similar effect; for whenever it was made, the whole army returned a hiss, and obeyed the signal by increasing their pace with the utmost hurry. The soldiers who had mounted aloft, and gave these signals, sat quite still during the interval (except making now and then a slight turn of the head), and seemed as solicitous to keep their posts as regular sentinels. The two columns of the army joined into one about twelve or fifteen paces from their separation, having in no part been above three yards asunder, and then descended into the earth by two or three holes. They continued marching by him for above an hour that he stood admiring them, and seemed neither to increase nor diminish their numbers, the soldiers only excepted, who quitted the line of march, and placed themselves at different distances on each side of the two columns; for they appeared much more numerous before he quitted the spot. Not expecting to see any change in their march, and being pinched for time, the tide being nearly up, and his departure being fixed at high-water, he quitted the scene with some regret, as the observation of a day or two might have afforded him the opportunity of exploring the reason and necessity of their marching with such expedition, as well as of discovering their chief settlement, which is probably built in the same manner as the large hills before described. If so, it may be larger and more curious, as these insects were at least one-third larger than the other species, and consequently their buildings must be more wonderful, if possible; thus much is certain, there must be some fixed place for their king and queen, and the young ones. Of these species he did not see the perfect insect.

In fine, although the curious and interesting habits and manners which have been here described have been well ascertained and accurately observed, naturalists are not yet agreed as to the true physiological characters of the most numerous of the classes composing these communities. That the two individuals called the king and queen in the preceding pages, are perfect insects, deprived of their wings, seems to be on all hands admitted; and that they are kept for the special purpose of propagation, and honoured as the common parents, is also certain. But the true character of the multitude of workers and soldiers is not so clear. Latreille inferred that the workers of Smeathman consist of the larvæ and pupæ, which later pass into the perfect state, assuming wings, and swarm in the manner already described; and that the soldiers are an order apart, which never assume the perfect state, and are incapable of reproduction. To this, Burmeister objects, that there is no instance in the whole animal world in which the undeveloped young labour for the old; and therefore doubts that the workers can be larvæ or pupæ; to which may be added, that these so-called larvæ still retain their form when the winged individuals appear. Huber also doubts that the soldiers can be properly called neuters, and Kirby thinks they are probably male larvæ. Westwood suggests that the soldiers as well as the workers remain wingless without changing their form, their development stopping short before arriving at maturity, and thereby some individuals acquire that enlarged head which distinguishes the soldiers, and that the real larvæ of the comparatively few specimens which ultimately become winged, are as yet unknown.

These vague and discordant conjectures of naturalists so eminent, show how much still remains to be discovered of the physiology of the White Ants.

INSTINCT AND INTELLIGENCE.

CHAPTER I.

1. Instinct defined. — 2. Independent of experience or practice. — 3. Sometimes directed by appetite.—4. A simple faculty independent of memory.—5. Instinctive distinguished from intelligent acts.—6. Instinct and intelligence always co-exist.—7. The proportion of instinct to intelligence increases as we descend in the organic chain.—8. Opinions of Descartes and Buffon—Character of the dog.—9. Researches and observations of Frederic Cuvier.—10. Causes of the errors of Descartes, Buffon, Leroy, and Condillac.—11. Degrees of intelligence observed in different orders of animals.—12. Accordance of this with their cerebral development.—13. Opposition between intelligence and instinct.—14. Consequences of defining their limits.—15. Example of instinct in ducklings.—16. In the construction of honeycomb.—17. The snares of the ant-lion.—18. Their mode of construction and use.—19. Spiders' nets.—20. Fishes catching insects.—21. Provident economy of the squirrel.—22. Haymaking by the Siberian lagomys.—23. Habitations constructed by animals.—24. The house of the hamster.—25. The habitation of the mygale, with its door.—26. Habitations of caterpillars.—27. Clothing of the larva of the moth.—28. Dwellings of animals which are torpid at certain seasons.—29. The Alpine marmot—Curious structure of their habitations.—30. Method of constructing them. — 31. Singular habits of these animals.—32. Instincts of migration.—33. Irregular and occasional migration.—34. General assembly preparatory to migration.—35. Occasional migration of monkeys.

1. In contemplating the habits and manners of animals, numerous acts are observed bearing marks of more intelligence and foresight than it is possible to suppose such agents to exercise. Since intelligence, therefore, cannot be admitted as the exciting cause for such actions, they have been ascribed to another power, called INSTINCT, which is defined to be one by which, independent of all instruction or experience, animals are unerringly directed to do spontaneously whatever is necessary for their preservation and the continuance of their species.

2. Instinct, therefore, must be regarded as a simple power or disposition emanating directly from the Creator, and producing its effects, without the intervention of any mental process. These effects, moreover, are susceptible of no modification by experience or repetition. A purely instinctive act is performed with as much facility and perfection at the first attempt as after repetition, no matter how long continued. The new-born infant seizes the mother's breast with its lips, draws the milk from it, and swallows that nourishing fluid—a very complicated physical process—as readily and as perfectly as it does after the daily experience and practice of ten or twelve months. The young bee just emerged from the cell, sets about the highly geometrical process of constructing its complicated hexagonal comb, and accomplishes its work with as much facility and perfection as the oldest inhabitant of the hive.

3. Instinct operates sometimes, but not invariably, by the intervention of physical appetite. Thus animals seek food, and the union of the sexes, not with the purposes which Nature designs to attain by these acts, but for the mere pleasure attending the gratification of appetite and passion. This pleasure is the bait which the Creator throws out to allure them to do what is indispensable for the preservation of the individual and the continuance of the species.

Thus, although animals seek food to satisfy hunger, the act is still instinctive. In the choice of food, that which is hurtful or poisonous is avoided, and that which is nutritious selected. The food which is suitable to the organs of digestion is always that to which the animal directs itself. These organs in some are adapted to vegetable, in others to animal food, and each species accordingly seeks the one or the other. Since it cannot be imagined that these animals are endowed with intelligence by which they are enabled to judge of the qualities of this or that species of aliment, it is clearly necessary to ascribe their acts in choosing always those which are suitable to them, to a power different from and independent of intelligence.

4. While instinct is a simple power, prompting acts apparently
114

the most complicated, and producing its effects at once in the most perfect manner and without any internal effort on the part of the agent, intelligence, on the contrary, is a faculty consisting of various distinct operations depending on experience and susceptible of indefinite improvement by exercise. The perceptions received from external objects are the data upon which it is exercised. These perceptions are capable of being revived and identified by the faculty called memory. Thus, having once perceived any given object, it is identified upon its recurrence by the consciousness that the perception it produces is the same as that which was formerly produced by it. Thus, objects once seen are known when seen again.

Memory is essential to almost all other acts of intelligence, the most simple of which is that by which the mind infers that any effect which has been once produced will be again reproduced by the same agent under like circumstances ; and the oftener such effects are observed to be reproduced, the more strong is the conviction that they will reappear.

5. Instinctive acts are done without any perception or consciousness of their consequences on the part of the agent. Intelligent acts, on the contrary, are performed not only with a consciousness of their consequences, but *because* of that consciousness. They are performed precisely with a view to produce the effects which are known by previous experience to have resulted from them.

6. It must not be supposed that instinct and intelligence cannot coexist, or that the animal endowed with either is necessarily deprived of the other. It is certain, on the contrary, that most animals are more or less gifted with both. In man, constituting the highest link in the chain of animal organisation, the faculty of intelligence predominates in an immense proportion over that of instinct. In passing to the next link, the relation between these faculties undergoes a change so enormous, that naturalists have regarded man not merely as a species of animal, but as an order of organised beings apart, being the sole genus of his order and the sole species of his genus.

7. In descending from link to link downwards along the chain of animal organisation, the play of intelligence is observed to bear a less and less proportion to that of instinct, until we arrive at the last links, where all trace of intelligence is lost, and animal life becomes a mere system of phenomena produced by instinctive impulses.

8. The question of the relative provinces and play of instinct and intelligence in the animal world, has been agitated among philosophers and naturalists from the earliest epochs down to our

own times. Descartes maintained that the inferior animals were mere automata, but that being constructed by Nature, they are incomparably more perfect than any which could be constructed by man. Buffon allowed them sensations, and a consciousness of present existence, but denied them all exercise of thought, reflection, the consciousness of past existence or memory, and the power of comparing their sensations or having ideas. Yet notwithstanding this, in other parts of his works, he admits that a power of memory, active, extensive, and retentive, cannot be denied to certain species. Thus, in his history of the dog, he says that an ardent, choleric, and even ferocious disposition, which renders that animal in the wild state formidable to all around it, gives place in the domestic dog to the most gentle sentiments, the most lively attachments, and the strongest desire to please. The dog, creeping to the feet of its master, places at his disposition its courage, its force, and its talents. It waits his orders merely to execute them ; it consults him, interrogates him, supplicates him, understands the slightest signs of his wishes : has all the warmth of sentiment which characterises man, without the light of his reason ; has more fidelity, more constancy ; no ambition, no selfish interest, no desire of vengeance, no fear save that of its master's displeasure. It is all zeal, all ardour, all obedience. More sensible to the memory of kindness than of injury, it is not disheartened by bad treatment. It submits and forgets, or remembers only the more to attach itself. Far from being irritated by, or flying from him who punishes it, it willingly exposes itself to new trials. It licks the hand which strikes it, offers no remonstrance save the expression of its pain, and disarms the hand which punishes it by patience and submission.*

Thus while Buffon refuses thought to the dog, he admits that he is capable of consulting, interrogating, and supplicating his master, and understanding the signs of his will. But, how, it may be asked, can a dog understand, without understanding ? Without the faculty of memory, how can he remember kindness and forget ill-treatment ? Buffon, as M. Flourens justly observes, admits as an historian, but he denies as a philosopher, and in spite of his acute understanding, allows his judgment to be influenced by the purpose to which the work on which he is engaged at the moment is directed. As an historian, he has to state facts ; and he does so with truth and eloquence. As a philosopher, he has to defend a system ; and he closes his eyes on all facts save those which support his hypothesis.

9. During more than a century which elapsed between the

---

* "Histoire du Chien," vol. 5, p. 196

epochs of Descartes and Buffon,* the question of the instinct and intelligence of animals was discussed in the spirit of the ancient philosophy on purely metaphysical grounds. It was with Buffon, and soon afterwards with Leroy, that it began to be placed upon the basis of observation and induction; but the first philosopher who reduced it to a definite form and supported his reasoning by observations systematically pursued was Frederick Cuvier. He proposed to determine the limits of the intelligence of different species; those which separate intelligence generally from instinct; and those in fine by which human intelligence is distinguished from that of inferior animals. These three points being once established, the long vexed question of animal intelligence was presented under a new aspect.

10. When Descartes and Buffon refused intelligence to animals, they did so because they could not accord to them the same faculty of intelligence which characterises the human race. Their error therefore arose from not perceiving or not defining the limit which separates human from animal intelligence.

When Condillac and Leroy, on the contrary, falling into the other extreme, accorded to animals the most elevated intellectual powers, they did so because they overlooked the distinction between instinct and intelligence. When they ascribed to intelligence acts which were prompted by instinct, and therefore executed with a perfection which, if they were the result of intelligence, would require a very elevated degree of that faculty, they were forced to admit in animals the possession of powers in some respects even more elevated than those of the human race.

11. The first observations of Frederick Cuvier indicated the various degrees of intelligence in the different orders of mammifers. Thus he found the highest development of that faculty in the *Quadrumana*, at the head of which stand the chimpanzee and ourang-outang. The second rank was assigned to the *Carnivora*, at the head of which was placed the dog. The *Pachydermata* stand next, with the horse and the elephant at their head; the two lowest ranks consisting of the *Ruminants* and *Rodents*.

12. Now it is important to remark that this classification of mammifers according to their relative intelligence, based upon the direct observation of their manners and habits, is found to be in complete accordance with their cerebral development; the organs of the brain, which in man have been ascertained as being those on which the intellectual functions depend, existing in a less and less state of development as we descend from the Quadrumana to the Carnivora, from the latter to the Pachydermata, and from these successively to the Ruminants and Rodents.

* Descartes published his "Discours sur la Méthode" in 1637; and Buffon published the "Discours sur la Nature des Animaux" in 1753.

The reader will find these conclusions verified by many of the examples which will be presently produced, but those who desire a more complete demonstration must have recourse to the numerous and beautiful memoirs of Frederick Cuvier, in which the original observations are recorded.

13. After having established the limits which distinguish the degrees of intelligence of different orders of animals, Cuvier took up the still more important question to fix the limit between intelligence and instinct.

Between these powers there is the most complete opposition. All the results of instinct are blind, necessary, and invariable. All those of intelligence, on the contrary, are optional, conditional, and susceptible of endless modification. The beaver, which builds its hut, and the bird which constructs its nest, act by instinct alone. The dog and the horse, which are educated so as to understand the signification of several words uttered by those who have charge of them, do so by the exercise of intelligence.

All the results of instinct are innate. The beaver builds its hut without having learned to do so. It is urged by a constant and irresistible force. It builds because it cannot help building.

All the results of intelligence arise from experience and instruction. The dog obeys his master, only because he has learned to do so. He is a free agent, and obeys because he wills to obey.

In fine, the results of instinct are particular, while those of intelligence are general. The industry and ingenuity which has excited so much admiration in the beaver, is displayed in nothing except the construction of his hut, while the same degree of attention and thought, which enables the dog to obey his master in one thing, will equally avail him to perform other acts.

14. So long as these two powers of instinct and intelligence were undistinguished one from the other, the manners and habits of animals presented to the contemplation of the observer endless obscurity, and the most perplexing contradiction. While in most actions the superiority of man over other animals is apparent, in many the superiority seems to pass to the side of the brute. This paradox and apparent contradiction disappears, however, when the boundary between instinct and intelligence is clearly marked. Whatever proceeds from intelligence in the lower animals, is incomparably below that which results from the intelligence of man ; and on the contrary, all those acts of the lower animals, which, supposing them to result from intelligence, would require a higher degree of that faculty than man possesses, are the mere effects of the blind mechanical power of instinct.*

* Flourens' " De l'Instinct et de l'Intelligence des Animaux," p. 36.
118

15. As an example of an act manifestly instinctive, a fact familiar to all who have visited a poultry-yard may be mentioned. When a mixed brood of chickens and ducklings hatched by a hen approach for the first time a pond of water, the ducklings precipitate themselves into the liquid, in spite of the efforts of their adopted mother to prevent them, and contrary to the example of the chickens, with whom they have come into life and from whom they have never been separated. The ducklings who do this may have never before seen water or any individuals of their own species, yet they use their webbed feet as propellers with as much skill as the oldest and most experienced of their race.

16. An example of a much more complicated process, which is manifestly instinctive, is presented by the labours of the bee already mentioned. The comb is a highly geometrical structure, which, if executed under the direction of intelligence, would require not only faculties of a high order, but profound calculation and much experience. Considered in relation to the purposes it is destined to fulfil, it would require the greatest foresight and a thorough knowledge of the whole course of life and organic functions of the species to which the constructors belong. Supposing them to be endowed with the necessary intelligence, the combs could not be constructed without many preliminary trials and partial failures, the necessary perfection being only attainable by slow degrees and by means of a series of experiments. Nothing of the kind however takes place, the complicated structure being produced at once with the greatest facility and in the highest perfection. There are, therefore, here none of the characters of a work directed by intelligence, but all the marks of one prompted by instinct.

17. Although the acts by which animals obtain and select their proper food are undoubtedly instinctive, they are, nevertheless, often attended with circumstances which it would be difficult to explain without the intervention of some degree of intelligence.

Fig. 1.—The Ant-Lion.

Fig. 2.—Larva of the Ant-Lion.

There is a little insect of the order *Neuroptera* and the family

Myrmeleonidæ, commonly called the *ant-lion,* represented in its natural size in fig. 1, the larva of which is also represented in its natural size in fig. 2. This larva feeds upon ants and other insects, of which it sucks the juice ; but as its powers of locomotion are greatly inferior to those of its prey, it would perish for want of nourishment, if Nature had not endowed it with instinctive faculties by which it is enabled to capture by stratagem the animals upon which it feeds.

18. After having carefully surveyed the ground upon which it is about to operate, it commences by tracing a circle corresponding in magnitude with its intended snare. Then placing itself within this circle, and using one of its feet as a spade or shovel, it sets about making an excavation with a tunnel-shaped mouth. It throws upon its head the grains of sand which are digged up with its feet, and by a jerk of its body it flings them to a distance of some inches outside the circle which it has traced, throwing them backwards by a sudden upward movement of the head. Proceeding in this way it moves backwards, following a spiral course, continually approaching nearer to the centre. At length so much of the sand is thrown out that a conical pit is formed, in the bottom of which it conceals itself, its mandibles being the only parts which it allows to appear above the surface. If in the course of its work it happens to encounter a stone, the presence of which would spoil the form of the pitfall, it first pays no attention to it, and goes on with its labour. After having finished the excavation, however, it returns to the stone, and uses every effort to detach it, to place it on its back and throw it out of the pit. If it do not succeed it abandons the work, and departs in search of another locality, where it recommences with admirable patience a similar excavation.

These pitfalls, fig. 3, when completed, are generally about three inches in diameter and two in depth ; and when the slope of the sides has been deranged, — which almost always happens when an insect falls into it,—the ant-lion immediately sets about repairing the damage.

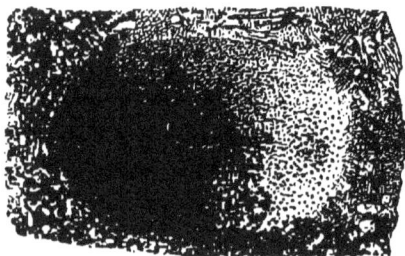

Fig. 3 —Pitfall of an Ant-Lion.

When an insect happens to fall into the pit, the ant-lion instantly seizes it and puts it to death, and the fluids having been all sucked out, its dry carcass is treated exactly like the grains of sand, and jerked out of the hole. If, however, as often happens, an insect who has the misfortune to fall from the brink of the precipice should recover itself, and escaping the murderous jaws of its enemy regain the summit, the latter immediately begins to throw up more sand, whereby not only is the hole made deeper, but its sides are rendered more precipitous, and the flying insect is often hit by the masses thus projected, and brought down again to the bottom.

19. Certain spiders spread snares still more singular. The web which these animals spread is destined to catch the flies and other insects upon which they prey. The disposition of the filaments composing this web varies with different species, but is often of extreme elegance.

20. There are certain fishes which feed upon insects that are not inhabitants of the water, and who resort to expedients, bearing marks of great ingenuity, to capture their prey. Thus, a species called the *Archer*, which inhabits the Ganges, feeds on insects which are accustomed to light upon the leaves of aquatic plants. The fish, upon seeing them, projects drops of water upon them with such sure aim, that it seldom fails to make them fall from the leaf into the water, when it seizes upon them. As the near approach of the fish would alarm the insect and cause its flight, this species of liquid projectile is usually launched from a distance of several feet, where the insect cannot see its enemy.

21. Certain species feed upon natural products, which are only to be found at particular seasons of the year; and in all such cases Nature prompts them, during their proper harvest, to collect and store up such a quantity of food as may be sufficient for their support, until the ensuing season brings a fresh supply. The common squirrel (fig. 4.) presents an example of this instinct. During the summer these active little creatures collect a mass of nuts, acorns, almonds, and other similar products, and establish their storehouse usually in the cavity of a

Fig. 4.—The Common Squirrel.

tree. They have the habit of providing several of these magazines in different hiding-places cunningly selected; and in winter, when the scarce season arrives, they never fail to find their stores, even when they are overlaid with snow. It is remarkable that this impulse to hide their food does not cease with the necessity for it, for they take the same care of the residue unconsumed upon the return of the ensuing season.

22. Another rodent, called by naturalists the *Lagomys pica*, which bears a close resemblance to the common rabbit, and inhabits Siberia, is endowed with an instinct still more remarkable, since it not only collects in autumn the herbage necessary for its sustenance during the long winter of that inhospitable country, but it actually makes hay exactly as do our agriculturalists. Having cut the richest and most succulent herbs of the field, it spreads them out to dry in the sun; and this operation finished, it forms them into cocks or ricks, taking care so to place them that they shall be in shelter from the rain and snow. It then sets about excavating a tunnel leading from its own hole to the bottom of these ricks, so that it may have a subterranean communication between its dwelling and its hayyard; taking care, moreover, that, the hay being gradually cut from the interior of each stack, the protection provided by the thatching of the external surface will not be disturbed.

23. Another form of that particular instinct the object of which is the preservation of the individual, is manifested in the art, with which certain species construct for themselves a suitable dwelling. In executing all the operations, often very complicated, directed to this purpose, their labours are invariably marked by the same general routine; although the operative by whom the work is executed has never before witnessed a similar process, and is aided by neither direction, plan, nor model.

We have already mentioned the structure of the honeycomb as an example of this, but the insect world abounds with others not less interesting.

The silkworm constructs for itself, with the delicate threads which it spins, a cocoon, in which it encloses itself, to undergo in safety its metamorphosis and to become a butterfly. The rabbit, in like manner, burrows for itself a dwelling, and the beaver constructs those little houses which have rendered it so celebrated. We shall, on another occasion, return to architectural instinct, in noticing the labours executed in common by animals which live in societies.

24. The hamster (fig. 5) is a little animal of the class of rodents, bearing a close resemblance to the common rat. It inhabits the fields throughout Europe and Asia, and inflicts much injury on the farmer and agriculturalist. This animal constructs for itself

a subterranean house, consisting of several rooms connected together by corridors. The dwelling has two communications with the surface, one consisting of a vertical shaft, by which the animal makes its entrances and exits; the other is an inclined shaft, merely used for the purposes of construction, the animal extruding through it the earth excavated

Fig. 5.—The Hamster.

in the formation of its habitation. One of the rooms is furnished, as the bedroom of the owner, with a couch of clean, dry grass, and is otherwise neatly kept. The others are used as store-rooms for the winter stock of provisions, which are amassed there in considerable quantities.

The form of the store-rooms is nearly spherical, and their diameter from 8 to 10 inches.

The female, who never lodges with the male, usually provides several of these vertical entrances to her habitation, with a view to give easy means of entrance to her young, when they are pursued by any enemy, and obliged precipitately to take refuge in their dwelling.

The number of store-rooms which they provide being determined by their stock of provisions, they are excavated in succession, when one is filled the animal beginning to make another.

The room which the female constructs as a nursery for her young ones never includes provisions. She brings there straw and hay to make beds for them. Two or three times a year she has five or six younglings, which she nurses for about six weeks, at which age she banishes them from her dwelling to provide for themselves. The depth of the dwelling varies with the age of the animal, the youngest making it at about the depth of a foot. Each successive year the depth is increased, so that the vertical shaft leading to the den of the old hamster often has a depth of more than five feet, the whole habitation, including dwelling-rooms, store-rooms, and communicating corridors, occupying a space having a diameter of 10 or 12 feet.

25. Certain spiders, known to zoologists by the name of *mygales,* execute works similar to those of the hamster, but much more complicated, for not only do they construct a vast and commodious habitation, but they place at its entrance a *door,* mounted upon

123

hinges (fig. 6). For this purpose the animal excavates in the ground a sort of cylindrical shaft three or four inches deep, and coats its sides with a tenacious plaster. It then fabricates a door, by uniting alternately layers of plaster and vegetable filaments. This trap-door is made exactly to fit the mouth of the shaft, to which it is hinged by cementing some projecting filaments against the upper edge of the plastered surface. The external surface of this trap-door is rough, and in its general appearance differs little from the surrounding ground. The inside surface, however, is smooth and nicely finished. On the side opposite to the hinge there is a row of little holes, in which the animal introduces its claws to bolt the door when any external enemy seeks to force it open.

Fig. 6.—Nest of the Mygale.

26. It is, however, among the countless species of insects that we find the most curious and interesting processes adopted for the construction of habitations. Many species of caterpillars construct houses by rolling up leaves and tying them together by threads spun by the animal itself. In the gardens, nests of this kind are everywhere to be seen, attached to the leaves of flowers and bushes. It is in this way that the caterpillar of the nocturnal butterfly, the *Tortrix viridana*, forms its nest (fig. 7).

27. Other insects construct habitations for themselves with the filaments of woollen stuff, in which they gnaw holes. Among these is the well-known larva of the common moth, popularly miscalled a worm, which is found to be so destructive to articles of furniture and clothing. With the woolly filaments which it thus cuts from the cloth, the caterpillar constructs a tube or sheath, which it continually lengthens as it grows. When it finds itself becoming too bulky to be at ease in this dwelling, it cuts it open along the side, and inserts a piece, by which its capacity is increased.

Fig. 7.—Nest of Tortrix Viridana.

28. Certain animals, which pass the cold season in a state of lethargy, not only prepare for themselves a suitable retreat, and a soft and comfortable bed, but when they become sensible of the

124

drowsiness which precedes the commencement of their periodical sleep, they take care to stop up the door of their house, as if they could foresee that a long interval must elapse before they shall want to go out, and that the open door would not only expose them to cold, but might give admission to dangerous enemies.

29. The alpine marmots supply examples of these curious manners.

Fig. 8.—Alpine Marmot.

These animals usually establish their dwellings upon the face of steep acclivities, which look to the south or the east; they assemble in large numbers for the excavation of these dwellings by their common labour. The form of their dens is that of the letter Y placed on its side, thus ⋈, the tail being horizontal, and one of the two branches being inclined upwards, and the other downwards. The cavity, which forms the tail of the Y, is the dwelling-room. It is carpeted with moss and hay, of which the animal makes an ample provision in summer. The upward branch leads to the door of the dwelling, and supplies the means of exit and entrance to the inhabitants. The descending branch is used for the discharge of ordure, and all other offal, the removal of which is necessary to the cleanliness of the house.

30. Buffon says, that in the construction of these dwellings, the animals observe a curious division of labour: some cut the grass, others collect it in heaps, and others, lying on their backs with their legs upwards, convert themselves into a sort of sledge, upon which the grass is heaped by the others, being kept together by the upright legs of the prostrate animal, just as hay is retained upon a farm-cart by the poles fixed at its corners. The animal lying thus is dragged by the tail by the others, to the mouth of the dwelling in which the grass is deposited.

The latter part of this statement is however called in question by some naturalists.

31. The marmots pass the greater part of their lives in these dens. They remain there during the night and generally during bad weather, coming out only on fine days, and even then not departing far from their dwelling. While they are thus abroad feeding and playing upon the grass, one of the troop, posted on a neighbouring rock, is charged with the duty of a sentinel, observing carefully the surrounding country. If he should perceive approaching danger, such as a hunter, a dog, or a bird of prey, he immediately gives notice by a long continued whistling or hissing noise, upon which the whole troop instantly rush to their hole.

Fig 9.—The White-throated Sajou.

32. There is another instinct worthy of notice, the object of which is always the preservation of the individual, and sometimes that of the species, which determines certain animals at particular epochs to undertake long voyages. These movements of migratory animals, as they are called, are sometimes periodic, being determined by the vicissitudes of the seasons, the animals being driven either from higher latitudes to lower by extreme cold, or from lower to higher by extreme heat. In other cases the migration is determined by the care of providing for its young; the animal migrating to localities where the food for its offspring abounds, and whence after depositing its eggs it departs

126

to places more conformable to its own habits and wants. Thus, the migration to and fro fulfils at once the double purpose of providing for the preservation of the species and that of the individual.

33. Where the migration is irregular, and the voyage not long, the movement is prompted by the necessity of seeking a locality where the proper nourishment of the animal is more abundant. In such cases, the animal having exhausted the supplies of a particular district, departs in quest of another, and does not voyage further than is necessary for that object.

Fig. 10.—The Maki.

34. Whatever be the motive which may prompt such voyages, they are almost invariably preceded by a general meeting, having all the appearance of a concerted one, composed of all the individuals of the species which inhabit the locality where it takes place. When the purpose of the voyage is change of climate, they do not wait until they are driven forth by an undue

127

temperature, but anticipate this change by an interval more or less considerable ; nor do they, as might be supposed probable, suffer themselves to be driven by degrees, from place to place, by the gradually increasing inclemency of the season. It would appear that they consider such a frequent change of habitation incompatible with their well-being, and instead of a succession of short voyages, they make at once a long one, which takes them into a climate from which they will not have occasion to remove until the arrival of the opposite season.

35. The monkeys, which abound in such vast numbers in the forests of South America, present an example of irregular migration. When they have devastated a district, they are seen in numerous bands bounding from branch to branch, in quest of another locality more abundant in the fruits which nourish them; and after the lapse of another interval, they are again seen in motion, the mothers carrying the young upon their backs and in their arms, and the whole troop giving itself up to the most noisy demonstrations of joy.

FIG. 19.—NEST OF THE GOLDFINCH.

INSTINCT AND INTELLIGENCE.

CHAPTER II.

36. Migration of the lemmings.—37. Vast migration of field-mice of Kamtschatka.—38. Instincts conservative of species stronger than those conservative of individuals.—39, 40. Instincts of insects for the preservation of their posthumous offspring.—41. 42. Transformations of insects—Precautions in the depositions of eggs.—43. Habitation constructed by liparis chrysorrhea for its young.—44. Examples mentioned by Reaumur and Degeer.—45. Expedients for the exclusion of light from the young.—46. Example of the common white butterfly. —47. Manœuvres of the gadfly to get its eggs into the horse's stomach.—48. The ichneumon.—49. Its use in preventing the undue multiplication of certain species.—50. Its form and habits.—51. The nourishment of its larvæ.—52. The sexton beetle.—53. Their processes in burying carcasses. — 54. Anecdote of them related by Strauss.—55. Singular anecdote of the gymnopleurus pilularius.— 56. Such acts indicate reasoning.—57. Anecdote of a sphex told by Darwin.—58. Indications of intelligence in this case.—59. Anecdote of a sexton beetle related by Gleditsch.—60. Indications of reason in this case.—61. Anecdote of ants related by Reaumur.—62. Anecdote of ants related by Dr. Franklin.—63. Anecdote of the bee related by Mr. Wailes. — 64. Anecdote of the humble bee by Huber. — 65.

36. IRREGULAR migrations, which are supposed to be in general determined by an instinctive presentiment of an approaching inclement season, are undertaken by small animals called lemmings, which have a close analogy to rats, and which inhabit

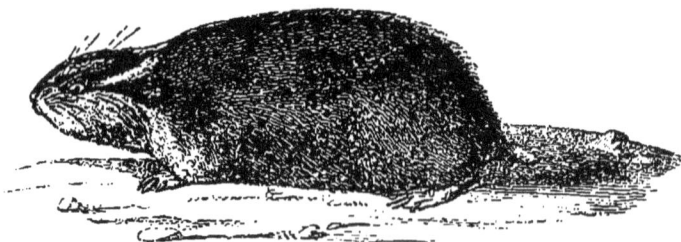

Fig. 11.—The Lemming.

the mountainous districts of Norway and the Frozen Ocean. These animals live in burrows, in which, like other similar species, they excavate rooms sufficiently spacious, in which they bring up their family. Their food consists in summer of herbs, and in winter of lichens. They lay up no store, however, and collect their supplies from day to day. By an inexplicable instinct, they have a fore-knowledge of a rigorous winter, during which the frozen ground would not allow them to collect their food in the country they inhabit. In such case, they emigrate in immense numbers, going to more favoured climates. This surprising presentiment of the character of the season has been frequently observed in this species. It was especially noticed in 1742. During that winter the season was one of extraordinary severity in the province of Umea, though much more mild in that of Lula, of which never-theless the latitude is higher. It was remarked, on this occasion, that the lemmings emigrated from the former province, but not from the latter.

On the occasions of such emigrations, countless numbers of troops of these animals, sometimes descending from the mountains, advance in close columns, always maintaining one direction, from which they never allow themselves to be turned by any obstacle, swimming across rivers wherever they encounter them, and skirting the rocks wherever they cannot climb over them. It is more especially during the night that these legions continue their march, reposing and feeding more generally during the day.

Although great numbers perish during the voyage, they nevertheless do immense damage to the districts over which they pass, destroying all the vegetation which lies in their way, and even turning up the ground, and consuming the fresh sown seed. Happily for the Lapland and Norwegian farmers, the visits of these animals are rare, seldom occurring more than once in ten years.

37. Such migrations, however, are much more frequently periodical, being determined, as already stated, by the change of seasons. Thus, it is found that in spring, immense legions of a little field-mouse, which inhabits Kamtschatka, depart from that country and direct their course towards the west. These animals, like the lemmings, proceeding constantly in one direction, travel for hundreds of leagues, and are so numerous that even after a journey of twenty-five degrees of longitude, in which a considerable proportion of their entire number must be lost, a single column often takes more than two hours to pass a given point. In the month of October they return to Kamtschatka, where their arrival constitutes a fête among the hunters, as they never fail to bring in their train a vast number of carnivorous animals, which supply furs in abundance to the inhabitants of these regions.

38. Nature seems even more sedulous for the preservation of the species than for that of the individual, and we find accordingly the instincts which are directed to the former purpose more strongly developed even than those of self-preservation. The animal world presents innumerable examples of this in the measures which nearly all species adopt with a view to the care of their young. The bird continues often for weeks to sacrifice all her own pleasures, and sits upon her eggs almost immovably. Before these eggs are laid she constructs with infinite labour and art a place in which she may with safety deposit them, and where the young which are destined to issue from them may be sheltered, protected, and fed by her until they have attained the growth and strength necessary to enable them to provide food and shelter for themselves.

39. The same instinct is manifested in a still more striking manner by insects. Many of these die immediately after they have laid their eggs, and consequently do not survive to see their young, of whose condition and wants therefore they can have no knowledge whatever by observation or experience. Their beneficent Maker has, however, taught them to provide as effectually for the security and well-being of their posthumous offspring, as if they had the most complete knowledge of their condition and wants. The effects of this instinct are so much the more remark-

able, as in many cases the young in their primitive state of larva inhabit an element and are nourished by substances totally different from those which are proper to their parent.

The instinct which guides certain animals to confer upon their young a sort of education, developing faculties and phenomena having a close analogy to those manifested in the conduct and operations of our own minds, never fails to excite as much astonishment as admiration, and teaches, more eloquently than words, how much above all that man can imagine or conceive, that power must be which has created so many wonders.

40. But the acts which manifest in the most striking manner the play of the instinctive faculty, are those already referred to by which insects, in the deposition of their eggs, adopt such precautions as are best calculated for the preservation of the young, which are destined to issue from these eggs when the provident mother is no more.

41. To comprehend fully this class of acts, it will be necessary to remind the reader that insects in general, before they attain their perfect state, pass through two preliminary stages, in which their habits, characters, and wants are totally different from those of the parent. The first stages into which the animal passes in emerging from the egg, is that of the *larva*, or grub; and the second, that of the *nymph*, or *pupa*.

Not only is the form and external organisation of the larva different from that of the insect into which it is destined to be ultimately transformed, but it is generally nourished by a different species of food, and often lives in a different element. Thus, while the perfect insect feeds upon vegetable juices, its larva is often voraciously carnivorous. While the perfect insect lives chiefly on the wing in the open air, the larva is sometimes aquatic, sometimes dwells on the hairs, or in the integuments, or even in the stomach or intestines of certain animals. The insect, therefore, cannot be imagined to know, from any experience, what will be the natural wants of the young which are destined to emerge from her eggs.

In many cases, any such knowledge on her part is still more inconceivable, inasmuch as the mother dies before her young break the shell. Nevertheless, in all cases, this mother, in the deposition of her eggs, is found to adopt all the measures which the most tender and provident solicitude for her young can suggest. If her young, for example, are aquatic, she deposits her eggs near the surface of water. If they are destined to feed upon the flesh or juices of any species of animal, she lays not only upon the particular animal in question, but precisely at those parts where the young shall be sure to find their proper nourishment.

132

If they are destined to feed upon vegetable substances, she deposits her eggs on the particular vegetables, and the particular parts of these vegetables which suit them. Thus, some insects lay their eggs upon the leaves of a certain tree, others in the bark of wood. Others again deposit them in the grain or seed of certain plants, and others in the kernel of certain fruits; each and all selecting precisely that which will afford suitable food to the larva when it breaks the shell.

42. But the care of the tender mother does not terminate here. As though she were aware that she will not herself be present to protect her offspring from the numerous enemies which will be ready to attack and devour it, she adopts the most ingenious expedients for its protection. With this view she envelops her eggs in coverings, which effectually conceal them from the view of the enemies to whose attack they would be exposed. In case the young should be susceptible of injury from the inclemency of the atmosphere, she wraps up the eggs in warm clothing, in which the young larva finds itself when it emerges from them.

43. Some species, such, for example, as the Liparis Chrysorrhea, envelop their eggs in a waterproof covering made of fur taken from their own bodies. They begin by forming with it a soft bed upon the surface of a branch, upon which they deposit several layers of eggs, which they then surround with more fur; and when all are laid, they cover them up with the same fur, the filaments of which, however, are differently disposed. The hairs which form the inside of the nest are arranged without much order, but, on the contrary, those which form its external covering are artfully arranged like the slates of a house, in such a manner that the rain which falls on them must glide off. When the mother has finished her work, which occupies her from twenty-four to forty-eight hours, her body, which before was invested with a clothing of rich velvet, is now altogether stripped, and she expires.

The females who thus provide for the protection of their young, often have the extremity of their bodies furnished with a great quantity of fur destined for this use.

44. Reaumur found one day a nest of this kind, but still more remarkable in its structure. The eggs were placed spirally round a branch, and covered with a thick and soft down, each hair of which was horizontal, which he described as resembling a fox's tail.

Degeer observed a proceeding, similar to those described above, with certain species of aphides, which cover their eggs with a cotton-like down, stripped from their own bodies by means of their hind-feet; but in this case the eggs were not enclosed in a common bed, but each in a separate covering.

45. These precautions seem to be intended not only to protect the eggs from wet and cold, but also to shade them from too strong a light, which would be fatal to the young they contain. It is doubtless for the same purpose that so many insects attach their eggs to the lower in preference to the upper surface of leaves, those which are placed on the upper surface being generally more or less opaque.*

46. The common white butterfly feeds upon the honey taken from the nectary of a flower, but her larva, less delicate and more voracious, devours the leaves of cabbage-plants. When we see, therefore, this insect flying about and alighting successively upon various plants, we imagine erroneously that she is in quest of her own food, when in reality she is searching for the plant whose leaves will form the proper nourishment for her future offspring. Having found this, and having carefully ascertained that it has not been pre-occupied by another of her species, she lays her eggs upon it and dies.

47. The young of the Gadfly (*Œstrus Equi*) are destined to live in the stomach of a horse. This being stated, it may well be asked how the insect fulfils that duty already described, which consists in depositing the eggs upon the very spot where the young will find their food ; for it can scarcely be imagined that the winged insect will fly down a horse's throat to lay in its stomach. Yet the parent accomplishes its object in a manner truly remarkable. Flying round the animal, she lights successively many times upon its coat, depositing several hundreds of her little eggs at the extremity of the hairs, to which she glues them by a liquid cement secreted in her body. This, however, would obviously fail to accomplish the purpose of supplying the young with their proper food, only to be found in the horse's stomach, to which, therefore, it is indispensable that the eggs should be transferred. Marvellous to relate, this transfer is made by the horse himself, who, licking the parts of his hide to which the eggs are attached, takes them, or the grubs evolved from them, if they have been already hatched, upon his tongue, and swallows them mixed with saliva ; thus conveying them to the only place where they can find their proper food !

But it may be objected, that by this process no eggs or grubs would find their way to the stomach, save those which might chance to be deposited upon those particular parts of the horse's body which it is accustomed to lick. There is, however, no chance in the affair ; for the insect, guided by an unerring and beneficent power, and as if foreseeing the inevitable loss of such of

* Lacordaire, Int. Ent., vol i., p. 29.

her young as might be deposited elsewhere, takes care to lay her eggs on those spots only, such as the knees and shoulders, which the horse is sure to lick!

48. Ichneumon was a name given to a certain species of quadrupeds, which were erroneously supposed to deposit their young upon the bodies of crocodiles, the entrails of which they gradually devoured. The name was transferred by Linnæus to a vast tribe of insects, whose young are destined to feed upon the living bodies of other insects, on which accordingly the mother deposits her eggs. The ichneumons were called by some naturalists *Muscæ vibrantes*, from the constant vibration of their antennæ, by which they were supposed, in some unknown manner, to acquire a knowledge of the insects which would be fit food for their young. This supposition is, however, clearly erroneous, inasmuch as many species do not manifest this vibratory motion.

49. The ichneumons are agents of vast importance in the economy of nature, by checking the too rapid increase of certain species, such as the caterpillars of butterflies and moths, of which they destroy vast numbers. The purpose of nature in this is un-mistakeably manifested by the fact, that the ichneumons increase in proportion to the increase of the species they are destined to destroy. Thus Nature maintains the equilibrium in the organic world as much by the operation of the destructive, as by that of the reproductive principle.

50. The ichneumon is a four-winged fly (fig. 12), which takes no other food than honey ; and the great object of the female is to discover a proper nidus for her eggs. In search of this she is in constant motion. Is the caterpillar of a butterfly or moth the appropriate food for her young? You see her alight upon the plants where they are most usually to be met with, run quickly over them, carefully examining every leaf, and having found the unfortunate object of her search, inserts her sting into its flesh, and there deposits an egg. In vain her victim,

Fig. 12.—The Ichneumon.

as if conscious of its fate, writhes its body, spits out an acid fluid, menaces with its tentacula, or brings into action the other organs of defence with which it is provided. The active ichneumon braves every danger, and does not desist until her courage and address have insured subsistence for one of her future progeny. Perhaps, however, she discovers, by a sense, the existence of which we perceive, though we have no conception of its nature,

135

that she has been forestalled by some precursor of her own tribe, that has already buried an egg in the caterpillar she is examining. In this case she leaves it, aware that it would not suffice for the support of two, and proceeds in search of some other yet unoccupied. The process is, of course, varied in the case of those minute species, of which several, sometimes as many as 150, can subsist on a single caterpillar. The ichneumon then repeats her operation, until she has darted into her victim the requisite number of eggs.

51. The larvæ hatched from the eggs thus ingeniously deposited, find a delicious banquet in the body of the caterpillar, which is sure eventually to fall a victim to their ravages. So accurately, however, is the supply of food proportioned to the demand, that this event does not take place until the young ichneumons have attained their full growth, when the caterpillar either dies, or, retaining just vitality enough to assume the pupa state, then finishes its existence; the pupa disclosing not a moth or a butterfly, but one or more full-grown ichneumons.

In this strange and apparently cruel operation one circumstance is truly remarkable. The larva of the ichneumon, though every day, perhaps for months, it gnaws the inside of the caterpillar, and though at last it has devoured almost every part of it except the skin and intestines, carefully all this time *avoids injuring the vital organs,* as if aware that its own existence depends on that of the insect on which it preys! Thus the caterpillar continues to eat, to digest, and to move, apparently little injured, to the last, and only perishes when the parasitic grub within it no longer requires its aid. What would be the impression which a similar instance amongst the race of quadrupeds would make upon us? If, for example, an animal—such as some impostors have pretended to carry within them—should be found to feed upon the inside of a dog, devouring only those parts not essential to life, while it cautiously left uninjured the heart, arteries, lungs, and intestines, should we not regard such an instance as a perfect prodigy, as an example of instinctive forbearance almost miraculous?[*]

Fig. 13.—The Sexton-Beetle.

52. The sexton-beetle, or Necrophorus (fig. 13), when about to deposit its eggs, takes care to bury with them the carcass of a mole or some other small quadruped; so that the young, which, like the

* Kirby, Int., vol. i., p. 288.

parent, feed upon carrion, the moment they come into existence, may have an abundant provision of nourishment.

53. The measures which these insects take to obtain and keep the carcasses upon which they feed, and which, as has been just observed, also constitute the food of their offspring, are very remarkable. No sooner is the carcass of any small dead animal discovered, such as a bird, a mole, or a mouse, than the sexton-beetles make their appearance around it to the number generally of five or six. They first carefully inspect it on every side, apparently

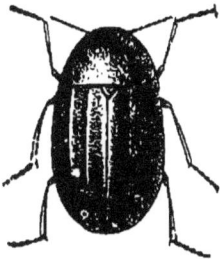

Fig. 14.—The Necrophorus Hydrophilus.　　Fig. 15.—The Marine Necrophorus.

for the purpose of ascertaining its dimensions, its position, and the nature of the ground on which it reposes. They then proceed to make an excavation under it, to accomplish which some partially raise the body, while others excavate the earth under the part thus elevated ; the operation being performed with the fore legs. By the continuance of this process, going round the body, they gradually make a grave under it, into which it sinks ; and so rapid is the process of excavation, that in a few hours the body is deposited in a hole ten or twelve inches deep. The males co-operate in this labour, and after it is accomplished, the female deposits her eggs upon the carcass.

54. Clarville * relates that he had seen one of these insects who desired thus to bury a dead mouse, but finding the ground upon which the carcass lay too hard to admit of excavation, it sought the nearest place where the soil was sufficiently loose for that purpose, and having made a grave of the necessary magnitude and depth, it returned to the carcase of the mouse, which it endeavoured to push towards the excavation ; but finding its strength insufficient and its efforts fruitless, it flew away. After some time it returned accompanied by four other beetles, who assisted it in rolling the mouse to the grave prepared for it.

* Cited by Strauss, Considérations Générales, p. 389.

55. A similar anecdote is related of a sub-genus of the Lamelli-cornes, called the *Gymnopleurus pilularius*, an insect which deposits its eggs in little balls of dung. One of these having formed such a ball, was rolling it to a convenient place, when it fell into a hole. After many fruitless efforts to get it out, the insect ran to an adjacent heap of dung, where several of its fellows were assembled, three of whom it persuaded to accompany it to the place of the accident. The four uniting their efforts, succeeded in raising the ball from the hole, and the three friends returned to their dunghill to continue their labours. *

56. It is difficult, if indeed it be possible, to explain acts like these by mere instinct, without the admission of at least some degree of the reasoning faculty, and some mode of intercommuni-cation serving the purpose of language. If such acts were com-mon to the whole species and of frequent recurrence, it might be possible to conceive them the results of the blind impulses of instinct; but being exceptional, and the results of individual accident, they are deprived of all the characters with which by common consent instinct is invested. On the contrary, there are many circumstances connected with this, which indicate a sur-prising degree of reason and reflection. Thus, when the insect goes to seek for assistance, it does not bring back, as it might do, from the swarm engaged on the dunghill, an unnecessary number of assistants. It appears to have ascertained by its own fruitless efforts how many of its fellows would be sufficient to raise the dung-ball. To so many and no more it imparts its distress and communicates its wishes; and how can it accomplish this unless we admit the existence of some species of signs, by which these creatures communicate one with another?

57. Darwin relates, that walking one day in his garden, he per-ceived upon one of the walks a sphex, which had just seized a fly almost as large as itself. Being unable to carry off the body whole, it cut off with its mandibles the head and the abdomen, only retaining the trunk, to which the wings were attached. With these it flew away; but the wind acting upon the wings of the fly, caused the sphex which bore it to be whirled round, and obstructed its flight. Thereupon the sphex again alighted upon the walk, and deliberately cut off first one wing and then the other, and then resumed its flight, carrying off its prey.

58. The signs of intelligence as distinguished from instinct are here unequivocal. Instinct might have impelled the sphex to cut off the wings of the fly before attempting to carry it to its nest, supposing the wings not to be its proper food; and if the head

* Illiger's Entomological Magazine, vol. i., p. 488.

and abdomen of all flies captured and killed by the sphex were cut off, the act might be explained by instinct. But when the fly is small enough to allow the sphex to carry it off whole, it does so, and it is only when it is too bulky and heavy that the ends of the body are cut off, for the obvious purpose of lightening the load. With respect to the wings, the detaching them was an after-thought, and a measure not contemplated until the inconvenience produced by their presence was felt. But here a most singular effort of a faculty to which we can give no other name than that of reason, was manifested. The progress of the sphex through the air was obstructed by the resistance produced by the wings of the fly which it carried. How is it conceivable that upon finding this, and not before, the sphex should suspend its progress, lay down its load, and cut off the wings which produced this resist-ance, if it did not possess some faculty by which it was enabled to connect the wings in particular, rather than any other part of the mutilated body of the fly, with the resistance which it encoun-tered, in the relation of cause and effect? To such a faculty I know no other name that can be given than that of reason, although I readily admit the difficulty of ascribing such an intellectual effort.

59. Gleditsch * relates that one of his friends desiring to dry the body of a toad, stuck it upon the end of a stick planted in the ground, to prevent it from being carried away by the sexton-beetle, which abounded in the place. This, however, was unavailing. The beetles having assembled round the stick, surveyed the object and tried the ground, deliberately applied themselves to make an excavation around the stick; and hav-ing undermined it, soon brought it to the ground, after which they not only buried the carcase of the toad, but also the stick itself.

60. Now this proceeding indicates a curious combination of circumstances which it appears impossible to explain without admitting the beetles to possess considerable reasoning power and even foresight. The expedient of undermining the stick can only be explained by their knowledge that it was supported in its upright position by the resistance of the earth in contact with it. They must have known, therefore, that by removing this support, the stick, and with it the toad, would fall. This being accom-plished, it may be admitted that instinct would impel them to bury the toad, but assuredly no instinct could be imagined to compel them to bury the stick; an act which could be prompted by no conceivable motive except that of concealing from those

* Phys. Bot. Œcon. Abhand., vol. iii., 220.

who might attempt to save the body of the toad from the attacks of the beetles, the place where it was deposited.

61. Among the innumerable proofs that animals are capable of comparing, and to a certain extent generalising their ideas, so as to deduce from them at least their more immediate consequences, and thereby to use experience as a guide of conduct, instead of instinct, Reaumur * mentions the case of ants, which being established near a bee-hive, fond as they are of honey, never attempt to approach it so long as it is inhabited; but if they happen to be near a deserted hive, they eagerly rush into it, and devour all the honey which remains there. How can we account for this abstinence from the inhabited hive, in spite of the strong appetite for its contents, so plainly manifested in the case of the empty one, if not by the knowledge that on some former occasion a rash attack upon an inhabited hive was visited by some terrific vengeance on the part of the bees?

62. Dr. Franklin was of opinion that ants could communicate their ideas to each other; in proof of which he related to Kalm, the Swedish traveller, the following fact. Having placed a pot containing treacle in a closet infested with ants, these insects found their way into it, and were feasting very heartily when he discovered them. He then shook them out, and suspended the pot by a string from the ceiling. By chance one ant remained, which, after eating its fill, with some difficulty found its way up the string, and thence reaching the ceiling, escaped by the wall to its nest. In less than half an hour a great company of ants sallied out of their holes, climbed the wall, passed along ceiling, crept along the string into the pot, and began to eat again. This they continued to do until the treacle was all consumed, one swarm running up the string while another passed down. It seems indisputable that the one ant had in this instance conveyed news of the booty to his comrades, who would not otherwise have at once directed their steps in a body to the only accessible route.†

63. A similar example of knowledge gained by experience, in the case of the hive-bee, is related by Mr. Wailes.‡ He observed that all the bees, on their first visit to the blossoms of a passion-flower (*Passiflora cærulea*) on the wall of his house, were for a considerable time puzzled by the numerous overwrapping rays of the nectary, and only after many trials, sometimes lasting two or three minutes, succeeded in finding the shortest way to the honey at the bottom of the calyx; but experience having taught them

---

* Mémoires, vol. ·., p. 709.
† Kirby and Spence, vol. ii., p. 422.
‡ Entomological Magazine, vol. i., p. 525.

this knowledge, they afterwards constantly proceeded at once to the most direct mode of obtaining the honey; so that he could always distinguish bees that had been old visitors of the flowers from new ones, the latter being at a loss how to proceed, while the former flew at once to their object.

Fig. 16.—The Humble Bee.

64. A similar fact is related of the humble bees by Huber,[*] who, when their bodies are too large to enter the corolla of a flower, cut a hole at its base with their mandibles, through which they insert the proboscis to extract the honey. If these insects adopted this expedient from the first, and invariably followed it, the act might be ascribed to instinct; but as they have recourse to it only after having vainly tried to introduce their body in the usual way into the opening of the corolla, it can scarcely be denied that they are guided by intelligence in the attainment of their end. The marks of experience, memory, and comparison, are unequivocal. When they find their efforts to enter the first flower to which they address themselves fruitless, they do not repeat them upon other flowers of the same sort, but directly attack the base of the corolla. Huber witnessed such proceedings repeatedly in the case of bean-blossoms.

65. Insects give proofs without number of the possession of the faculty of memory, without which it would be impossible to turn to account the results of experience. Thus, for example, each bee, on returning from its excursions, never fails to recognise its own hive, even though that hive should be surrounded by various others in all respects similar to it.

66. This recognition of home is so much the more marked by traces of intelligence rather than by those of instinct, inasmuch as it depends not on any character merely connected with the

* Philosophical Transactions, vol. vi., p. 222.

hive itself, whether external or internal, but from its relation to surrounding objects; just as we are guided to our own dwellings by the recollection of the particular features of the locality and neighbourhood. Nor is this faculty in the bee inferred from mere analogies; it has been established by direct experiment and observation. A hive being removed from a locality to which its inhabitants have become familiar, they are observed, upon the next day, before leaving for their usual labours, to fly around the hive in every direction, as if to observe the surrounding objects, and obtain a general acquaintance with their new neighbourhood.

67. The queen in like manner adopts the same precaution before she rises into the air, attended by her numerous admirers, for the purposes of fecundation.

68. This curious example of the memory of bees is beautifully noticed by Rogers, in his poem on that faculty.

> " Hark ! the bee winds her small but mellow horn,
> Blithe to salute the sunny smile of morn.
> O'er thymy downs she bends her busy course,
> And many a stream allures her to its source.
> 'Tis noon, 'tis night. That eye so finely wrought,
> Beyond the search of sense, the soar of thought,
> Now vainly asks the scenes she left behind ;
> Its orb so full, its vision so confined !
> Who guides the patient pilgrim to her cell ?
> Who bids her soul with conscious triumph swell ?
> With conscious truth retrace the mazy clue
> Of varied scents that charmed her as she flew ?
> Hail, MEMORY, hail ! thy universal reign
> Guards the least link of Being's glorious chain."

69. The poet, however, has fallen into an error, as often happens when poets derive their illustrations from physical science. The bee is not reconducted to its habitation by retracing the scents of the flowers it has visited; for, if it were, it is obvious that in returning it would necessarily follow the zig-zag and tortuous course from flower to flower which it had followed during the progress of its labours in collecting the sweets with which it is loaded; whereas, on the contrary, in its return, no matter what be the distance, it flies in a direct line to its hive.

70. Kirby mentions the following curious fact illustrating the memory of bees, which was communicated to him by Mr. William Stickney, of Ridgemont, Holderness.

About twenty years ago, a swarm from one of this gentleman's hives took possession of an opening beneath the tiles of his house, whence, after remaining a few hours, they were dislodged and hived. For many subsequent years, when the hives descended from this stock were about to swarm, a considerable party of scouts were observed for a few days before to be reconnoitring about the old hole under the tiles; and Mr. Stickney is persuaded that if suffered they would have established themselves there. He is certain that for eight years successively the descendants of the very stock that first took possession of the hole frequented it, as above stated, and *not* those of any other swarm; having constantly noticed them, and ascertained that they were bees from the original hive, by powdering them while about the tiles with yellow ochre, and watching their return. And even later there were still seen, every swarming season, about the tiles, bees which Mr. Stickney has no doubt were descendants from the original stock.

71. Among the instincts manifested by insects, there is none more remarkable or more admirable than that already mentioned, by which certain species provide a store of food for their young, which differs totally from their own aliment, and which they would themselves regard with disgust. The pompilides, a species resembling wasps, are endowed with this faculty. The insect in its adult state feeds, like the bee, upon floral juices. But its young, in the infant state of larva, is carnivorous. The provident mother, therefore, when she deposits her eggs, never fails to place beside each of them in the nest, in a place prepared to receive it, the carcase of a spider or of some caterpillar, which she has slain with her sting for that express purpose.

72. The carpenter bee presents another example of this remarkable instinct, boring with incredible labour in solid wood a habitation which, though altogether unsuitable to itself, is adapted with the most admirable fitness for its young. Among these, one of the most remarkable is the *Xylocopa violacea*, fig. 17, a large species,* a native of middle and southern Europe, distinguished by beautiful wings of a deep violet colour, and found commonly in gardens, where she makes her nest in the upright putrescent espaliers or vine-props, and occasionally in the

Fig. 17.—The Carpenter Bee.

garden-seats, doors, and window-shutters. In the beginning of spring, after repeated and careful surveys, she fixes upon a piece of wood suitable for her purpose, and with her strong mandibles

* Kirby, vol. i., p. 369.

begins the process of boring. First proceeding obliquely downwards, she soon points her course in a direction parallel with the sides of the wood, and at length, with unwearied exertion, forms a cylindrical hole or tunnel, not less than twelve or fifteen inches long and half an inch broad. Sometimes, where the diameter will admit of it, three or four of these pipes, nearly parallel with each other, are bored in the same piece.

Fig. 18.—Nest of the Carpenter Bee.

Herculean as this task, which is the labour of several days, appears, it is but a small part of what our industrious bee cheerfully undertakes. As yet she has completed but the shell of the destined habitation of her offspring; each of which, to the number of ten or twelve, will require a separate and distinct apartment. How, you will ask, is she to form these? With what materials can she construct the floors and ceilings? Why, truly God "doth instruct her to discretion and doth teach her."

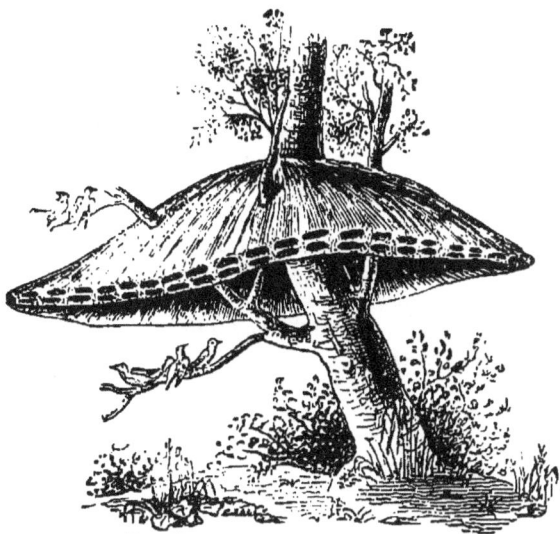

F:G. 23.—NESTS OF THE REPUBLICAN.

INSTINCT AND INTELLIGENCE.

CHAPTER III.

In excavating her tunnel, the carpenter bee has detached a large quantity of fibres, which lie on the ground like a heap of sawdust. This material supplies all her wants. Having deposited an egg at the bottom of the cylinder along with the requisite store of pollen and honey, she next, at the height of about three-quarters of an inch (which is the depth of each cell), constructs of particles of the sawdust, glued together, and also to the sides of the tunnel, what may be called an annular stage or scaffolding. When this is sufficiently hardened, its interior edge affords support for a second ring of the same materials, and thus the ceiling is gradually formed of these concentric circles, till there remains only a small orifice in its centre, which is also closed with a circular mass of agglutinated particles of sawdust. When this partition, which serves as the ceiling of the first cell and the flooring of the second, is finished, it is about the thickness of a crown piece, and exhibits the appearance of as many concentric circles as the animal has made pauses in her labour. One cell being finished, she proceeds to another, which she furnishes and completes in the same manner, and so on until she has divided her whole tunnel into ten or twelve apartments.

When the work here described is considered, it is evident that its execution must require a long period of hard labour. The several cells must be cut out, their floors agglutinated, and they must be each supplied with a store of honey and pollen, the collection and accumulation of which is a labour which must occupy a considerable interval of time ; and as the eggs are deposited successively in the cells according as they are finished and furnished, it is evident that they must be at any given moment in very different states of progress, the young issuing from those first deposited many days before the latest break the shell. But since there are ten or twelve such chambers vertically superposed, and since the lowest are the first laid, the new-born larva would either be condemned to be imprisoned in its cell until the births of all those above it should take place, or, in escaping to the exterior, it would have to pass through the chambers of all the others not yet developed, and would thus damage or destroy them. The beneficent Creator of the insect has, however, endowed it with an instinct which supplies the place of the foresight necessary to provide against such a catastrophe. With admirable forethought she constructs, besides the door already mentioned leading from cell to cell, another orifice in the lowest cell, which serves as a sort of postern, through which the insects produced from the earliest eggs emerge into day. In fact, all the young bees, even the uppermost, make their exit by this road ; for each grub, when about to pass into the state of pupa, places itself in its cell with its head downwards, and is thus necessitated, when arriving at the perfect state, to pass through the floor in that direction.*

73. It is especially in the first moment of their lives that animals in general are feeble, tender, and helpless, and have need of shelter from atmospheric vicissitudes, and protection from the attacks of their enemies; and we find, accordingly, that it is precisely these directions which have been given to the most irresistible instincts with which Almighty Goodness has endowed their parents. The number of species which in mature age build habitations for their own use, is insignificant compared with those which construct, with a labour which seems guided by the most touching tenderness and forethought, habitations for their young.

74. This habit is especially observable with birds. It is impossible to regard with sentiments other than those of the most profound interest the perseverance with which these creatures bring—straw by straw, and hair by hair—the materials destined for the formation of their nests, and the art with which they arrange them. The form, structure, and locality of these habitations is always the same for the same species, but different for different species, and are ever admirably adapted to the circumstances in which the young family are destined to live. Sometimes these cradles are constructed in the earth, and in a rude manner; sometimes they are cemented to the side of a rock, or to the wall of a building, but more commonly they are placed in the branches of trees, a hemispherical form being given to them (fig. 19.) They resemble, in form and structure, a little basket, rounded at the bottom and hollowed out at the top, the sides of which are formed of blades of grass, flexible straws and twigs, and hairs taken from the wool of animals, the inside being lined with moss or down.

Fig. 20.—Nest of the Baya.

---

* Reaur. vi. 39—52; Mon. Ap. Angl. i. 189; Apis. * * a. 2. β.

75. Sometimes, however, a much more complicated and artificial structure is produced. The nest of the baya, a little bird of India, resembling the bullfinch, (fig. 20,) has the form of a flask, and is suspended from some branch which is so flexible that neither serpents, monkeys, nor squirrels can approach it. But still more effectually to secure the safety of their young, the mother places the door of the nest at the bottom, where it can only be reached by flying. This habitation would be liable to fall to pieces if it were formed of straws or filaments laid horizontally; it is, therefore, constructed with admirable skill of blades or filaments arranged longitudinally. Internally it is divided into several chambers, the principal of which is occupied by the mother sitting on her eggs; in another the father of the family is accommodated, who is assiduous in his attentions to his companion, and while she fulfils with exemplary tenderness her maternal duties he amuses her with his song.

76. Another oriental bird, called the sylvia sutoria, or sewing wood-bird, builds a nest equally curious. This little creature, collecting cotton from the cotton-tree, spins it with its bill and claws into threads, with which it sews leaves round its nest, so as to conceal its young from their enemies (fig. 21).

77. Different species of animals are governed by social instincts which vary, but are always conducive either to the preservation or the well-being of the individual, or to the continuance of the race. When the food by which they are nourished is not so abundant as to support any considerable numbers in the same locality—which is generally the case with the larger species of carnivorous animals —they are endowed with an antisocial instinct, and not only lead a solitary life, but in many cases will not suffer any animal of their own species to remain in their neighbourhood.

78. Occasionally, however, the operation of this instinct is suspended. This takes place either when a scarcity of subsistence forces them to seek for food in places where they would be liable to attacks, against which their individual force would be insufficient for defence, or where some large flocks of animals of the sort on which they prey happen to come into their neighbourhood. In such cases they assemble by common consent in considerable numbers, and attack their prey in a body. Thus

Fig. 21.—Nest of the Sylvia Sutoria.

148

we see in the winter season bands of wolves, impelled by hunger, descend from the hills or forests and ravage the stock of the farmer,—an enterprise on which they never venture when other food can be obtained at less risk. In such cases, however, when the immediate object of their enterprise has been accomplished, their antisocial instinct revives, and they disperse, often quarrelling among themselves.

79. Various species which do not habitually live in society, nevertheless assemble in vast numbers, when at certain seasons they make long journeys. This is the case generally with migratory animals. The social instinct is, however, only temporary, since, when the journey is completed, and they arrive at their destination, they disperse.

80. The migratory pigeons of North America present a remarkable.example of this instinct, of temporary and periodical sociability. These birds, when stationary, are dispersed in vast numbers over the country, but when about to migrate, they assemble in inconceivable numbers, and perform their journey together, flying in a close and dense column nearly a mile in width, and six or eight miles in length. Wilson, the well-known American ornithologist, saw a flock of these birds pass over him in the state of Indiana, the number of which he estimated at two millions. The celebrated Audubon related that one day in autumn, having left his house at Henderson, on the banks of the Ohio, he was crossing an inclosed tract near Horsdensburgh, when he saw a flight of these pigeons, more than commonly numerous, directing their course from the north-west towards the south-east. As he approached Louisville, the flock became more and more numerous; he described its density and extent to be such, that the light of the sun at noon was intercepted, as it would have been by an eclipse, and that the dung fell in a thick shower like flakes of snow. Upon his arrival at Louisville, at sunset, having travelled fifty-five miles, the pigeons were still passing in dense files. In fine, this prodigious column continued to pass for three entire days, the whole population having risen and resorted to fire-arms to destroy them.

The usual habitations of these birds are the extensive woods which overspread that vast continent. A single flock will often occupy one entire forest ; and when they remain there some time, their dung is deposited on the ground in a stratum several inches thick. The trees are stript throughout an extent of many thousand acres, and sometimes completely killed, so that the traces of their visit are not effaced for many years.

81. Of all mammifers, the Canadian beaver is the most remarkable for sociability, industry, and foresight. During the summer

it lives alone in burrows, which it excavates on the borders of lakes and rivers ; but on the approach of winter, the animals quit these retreats, and assemble together for the purpose of constructing a common habitation for the winter season. It is in the most solitary places that they display their architectural instinct.

82. Two or three hundred having concerted together, select a lake or river too deep to be frozen to the bottom, for the establishment of their dwellings. They generally prefer a running stream to stagnant water, because of the advantage it affords them as a means of transport for the materials of their habitation. To keep the water at the desired depth, they commence by constructing a dam or weir in a curved form, the convexity being directed against the stream. This is constructed with twigs and branches, curiously interlaced, so as to form a sort of basket-work, the interstices being filled with gravel and mud, and the external surface plastered with a thick and solid coating of the same. This embankment, the width of which, at its base, is commonly from twelve to fourteen feet, lasts, when once constructed, from year to year, the same troop of beavers always returning to pass the winter under its shelter. Their labours after the first season are limited to keeping it in repair ; they strengthen it from time to time by new works, and restore whatever may be worn away by the action of the weather. It is rendered more permanent by a vigorous vegetation, which soon clothes its surface.

83. Wherever stagnant water has been selected, this preliminary labour becomes unnecessary, and the animals proceed at once to build their village. But, as has been already observed, they are subject in that case to an equivalent amount of labour in the transport of the materials.

When this preliminary work has been completed, they resolve themselves into a certain number of families, and if the locality is a new one, each family sets about the construction of its huts ; but if they return to the village they inhabited a former year, their labour is limited to the general repair and cleansing of the village.

The cabins composing it are erected against the dam, or upon the edge of the water, and generally have an oval form. Their internal diameter is six or seven feet, and their walls, like the dam, constructed of twigs and branches, are plastered on both sides with a thick coating of mud. The cabin, of which the foundation is below the surface of the water, consists of a basement and an upper storey, the latter being the habitation of the animals, and the former serving as storeroom for provisions.

The entrance to the cabin is in the basement story, and below the level of the water. .

150

It has been supposed that the animal uses its tail as a trowel in building these habitations. It appears, however, that this is an

Fig. 22.—The Beaver.

error, and that they use only their teeth and the paws of their fore-feet. They use their incisive teeth to cut the branches, and when necessary the trunks of trees; and it is with their mouth and their fore-feet that they drag these materials to the place where they intend to erect their habitation. When they have the advantage of running water, they take care to cut their wood at a point on the banks of the stream above the place where they are about to build. They then push the materials into the water, following and guiding them as they float down the stream, and landing them, in fine, at the point selected for their village. It is also with their feet that they excavate the foundations of their dwellings. These labours are executed with great rapidity and chiefly during the night.

84. The beaver, being a mammifer of the order of rodents, is one of the classes to which Cuvier assigns, as has been already stated, the lowest degree of intelligence. If the various acts here related were assigned to intelligence, they would evince a high degree of that faculty. Cuvier, however, demonstrated conclusively that they were acts altogether instinctive. He took several young beavers from their dams, and reared them altogether apart from their species, so that they had no means of acquiring any knowledge of the habits and manners of their kind. These animals, brought up in cages, isolated and solitary, where they had no natural necessity for building huts, nevertheless, pushed by the blind and mechanical force of instinct, availed themselves of materials, supplied to them for the purpose, to build huts.

151

85. In the low estimate of intelligence assigned by Cuvier to the beaver, other naturalists concur. "All agree," says Buffon, "that this animal, far from having an intelligence superior to others, as would necessarily be the case if his architectural skill were admitted to be the result of such a faculty, appears, on the contrary, to be below most others in its individual qualities. It is an animal, gentle, tranquil, familiar; of plaintive habits, without violent passions or strong appetites. When confined it is impatient to recover its liberty, gnawing from time to time the bars of its cage, but doing so without apparent rage or precipitation, and with the sole purpose of making an opening by which it may get out. It is indifferent; shows no disposition to attachment, and seeks neither to injure nor to please those around it. It seems made for neither obedience nor command, nor even to have commerce with its kind. The spirit of industry which it displays when assembled in troops, deserts it when solitary. It is deficient in cunning, without even enough of distrust to avoid the most obvious snares spread for it; and, far from attacking other animals, it has not the courage or skill to defend itself."

86. The pursuit of the beaver has been prosecuted to such an extent in Canada, that the animal has been nearly exterminated there, and more recently the trappers have been obliged to extend their excursions in search of them to the sources of the Arkansas, in the Rocky Mountains. The snare or trap used for catching the animal is similar to that used for foxes and polecats. The trappers, who make their excursions in caravans for mutual protection against the attacks of the Indians, acquire such skill, that they discern at a glance the track of the animal, and can even tell the number which occupy the hut. They then set their traps at a few inches below the surface of the water, and connect them by chains to the trunk of a tree, or to a stake planted strongly in the bank. The bait consists of a young twig of willow, stripped of its bark, the top rising to five or six inches above the surface of the water. The twig has been previously steeped in a sort of decoction made from the buds of poplar, mint, camphor, and sugar. The beaver, being gifted with a fine sense of smell, is attracted by the odour, and in touching the twig he disengages the detent of the trap and is caught.

87. The social instinct is not so common among birds as with mammifers, nevertheless some remarkable examples of it are found, among which may be mentioned a species of sparrow called the *republican*, which lives in numerous flocks in the neighbourhood of the Cape of Good Hope. These birds construct a roof (fig. 23), under which the whole colony build their nests.

88. But it is among insects we must look for the most striking manifestations of the architectural instinct.

The wasp (fig. 24.) affords an example of this, scarcely less interesting than the well-known economy of the bee. These little animals, though ferocious and cruel towards their fellow insects, are civilised and polished in their intercourse with each other, and compose a community whose architectural labours will not suffer by comparison even with those of the peaceful inhabitants of the hive. Like the latter, their efforts are directed to the erection of a structure for their beloved progeny, towards which they manifest the greatest tenderness and affection. They construct combs consisting of hexagonal cells for their reception; but the substance they use for this purpose is altogether different from wax, and their dwelling is laid out upon a plan in many respects different from that of the bee.

Fig. 24.—The Wasp.

89. Their community consists of males, females, and neuters. At the commencement of spring a pregnant female, which has survived the winter, commences the foundation of a colony destined before the autumn to become a population of some twenty or thirty thousand. The first offspring of this fruitful mother are the neuters, who immediately apply themselves to the task of constructing cells, and collecting food for the numerous members of the family who succeed them; and it is, while engaged in this labour, that they are most disposed to avenge themselves upon all who attempt to molest or interrupt them.

90. It is not till towards the autumn that the males and females are brought forth. The males as well as the neuter soon die, and the females surviving, seek some place of refuge in which to pass the winter, being previously impregnated.

91. The nest of the common wasp, generally built under ground, is of an oval form, from sixteen to eighteen inches high, and from twelve to thirteen in diameter.

Another species builds a nest of nearly the same form, but suspends it from the branches of trees; the size of these suspended nests varying from two inches to a foot in diameter. A section of the underground nest of a common wasp is shown in fig. 25.

It is a singular fact that the material of which the wasp builds its habitation is paper, an article fabricated by this insect ages before the method of making it was discovered by man.

With their strong mandibles they cut and tear from any pieces

153

of old wood to which they can find access, a quantity of the woody fibre, which they collect into a heap and moisten with

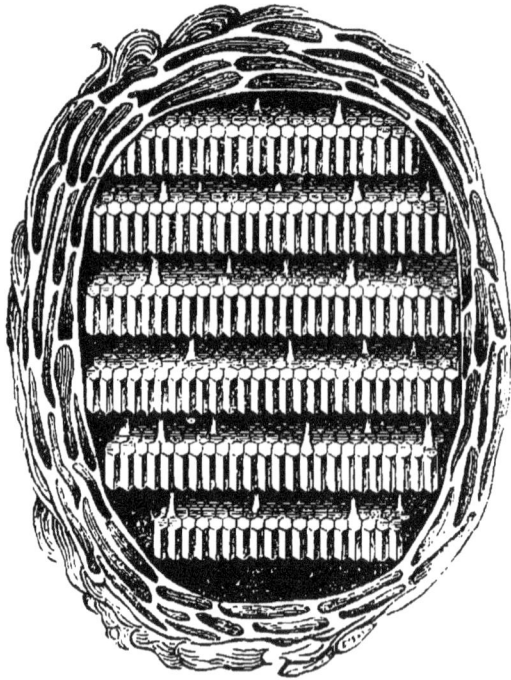

Fig. 25.—Underground Wasp's nest.

viscid liquid secreted in their mouths. They knead this with their jaws until they form it into a mass of pulp similar precisely to that which the paper-maker produces from the vegetable fibre of linen or cotton rags. With this pulp, they fly off to their nests, where, by walking backwards and forwards, they spread it out into leaves of the necessary thinness by means of their jaws, tongue, and legs. This operation is repeated many times, until at length as much of the paper is produced as is sufficient to roof in the nest. The thinness of this wasp-made paper is about the same as that of the book now in the hands of the reader.

The coating of the nest consists of fifteen or sixteen leaves of this paper placed one outside the other, with small spaces between them as shown in the figure, so that if rain should chance to penetrate one or two of them, its progress may be arrested by the inner ones.

92. The interior of the nest consists of from twelve to fifteen horizontal layers of comb placed one over the other so as to form

as many distinct and parallel storeys. And here we may observe in passing, the difference between the architectural system of the wasp and that of the bee. The latter builds its cells in vertical strata ranged side by side, the mouths opening horizontally so that the insects in passing between stratum and stratum must creep up the intervening vertical corridors; while the wasp, on the other hand, prefers horizontal corridors, so that in passing between stratum and stratum it creeps over one and under the other. In short, the positions given to the ranges of comb by the bee, in contradistinction to that adopted by the wasp, will be understood by supposing the sides of the wasp's habitation to represent the top and bottom of that of the bee.

Each comb of the wasp is composed, as shown in the figure, of a numerous assemblage of hexagonal cells made of the same paper as that already described, each cell being distinct, with double partition-walls. These cells, unlike those of the hive bee, are arranged only in a single row, the open end of each cell being turned downwards and the upper end being closed by a slightly convex lid, and not by a pyramidal cover like those of the honey-comb. The upper surface of each stratum of comb is therefore a continuous floor formed like an hexagonal mosaic, the surface being nearly but not perfectly smooth, since each hexagonal piece is curved slightly upwards.

The open mouths of the cells being presented downwards, the nurses as they creep along the roof of each stratum can easily feed the young grubs which occupy the cells of the stratum immediately above. The space left between one stratum and another is about half an inch.

Each stratum of comb is attached at the sides of the walls of the nest, but the tenacity of the paper of which the comb is composed would not be sufficient to sustain the weight of the stratum when the cells are all filled with grubs. The little architects, therefore, as though they had foreseen this, take care to connect at regulated intervals each stratum with that below it by strong cylindrical columns or pillars. Each of these, like the columns used in architecture, has a base and a capital, to which greater dimensions are given than those of the connecting shaft. These columns are composed of paper similar to that used for other parts of the nest, but of a more compact and stronger texture. The middle strata are connected by a colonnade of from forty to fifty of these pillars; the number being less as the dimensions of the strata decrease in going upwards or downwards.

93. The process of building this structure is as follows. The dome is first completed, as already described, by laying fifteen or

sixteen little sheets of paper one under the other, with intervening spaces at each part of it. Before the walls are further continued, the first or uppermost stratum of comb is then fabricated and attached to the sides by paper cement, and to the roof by a colonnade of pillars. The empty cells of this stratum being ready, the female big with eggs, deposits an egg in each, which is retained there by being agglutinated to the roof and sides of the cell: meanwhile, the workers continue their architectural labours, first carrying downwards the paper walls as already described, and next constructing the second stratum of comb and connecting it with the first by a colonnade.

94. It must be observed that in the society there is a well-organised division of labour. One part of it is employed exclusively in building, another in collecting food for the young, and in tending and nursing them, and, in fine, the female in depositing eggs in the cells. Since, therefore, a comparatively small proportion of the colony is engaged in building, the progress of the structure is necessarily slow, its entire completion being the work of several months; yet, though the result of such severe labour, it merely serves during the winter as the abode of a few benumbed females, and is entirely abandoned on the approach of the spring, wasps never using the same nest for more than a single season.[*]

95. The cells, which in a populous nest are not fewer than 16000, are of different sizes, corresponding to that of the three orders of individuals which compose the community; the largest for the grubs of females, the smallest for those of workers. The last always occupy an entire comb, while the cells of the males and females are often intermixed.

96. Besides openings which are left between the walls of the combs to admit of access from one to the other, there are at the bottom of each nest two holes, by one of which the wasps uniformly enter, and through the other issue from the nest, and thus avoid all confusion or interruption of their common labours.

97. As the nest is often a foot and a half under ground, it is requisite that a covered way should lead to its entrance. This is excavated by the wasps, who are excellent miners, and is often very long and tortuous, forming a beaten road to the subterraneous dwelling, well known to the inhabitants, though its entrance is concealed from incurious eyes. The cavity itself, which contains the nest, is either the abandoned habitation of moles or field mice, or a cavern purposely dug out by the wasps, which exert themselves with such industry as to accomplish the arduous undertaking in a few days.[†]

---

[*] Reaum. vi. 6. [†] Kirby, vol. i. p. 426.

98. While it is incontestable that instinct is the predominant spring of action with the inferior species, it is nevertheless impossible to deny many animals the possession of a certain degree of intelligence. Many are evidently endowed, not only with memory, but even with judgment, and a certain degree of the reasoning faculty.

99. That many species possess the faculty of memory in a high degree of development is evident. Domesticated animals in general know and remember their homes and their owners. A horse, even after having made a single excursion from his stable, will recognise the road to it on his return, and it is even affirmed that upon returning after several years' absence to a locality which he has inhabited for a sufficient time to become familiar with it, he will again recognise it, and left to himself will find his way into the stable he formerly occupied, and resume the possession of his former stall. The dog, the elephant, and other domesticated animals, recognise, even after longer intervals, those who have treated them well or ill, and manifest accordingly their gratitude or their vengeance.

100. It happened not long since that an elephant in one of the collections publicly exhibited in this country, extending his trunk between the bars of his stall, suddenly struck down with it an individual among a crowd of spectators, obviously selected by the animal for the infliction of the blow. A circumstance so singular excited inquiry, more especially as it was seen that the person attacked had not in any way at the time offended or molested the animal. It was ascertained, however, upon inquiry, that some weeks previously the same individual had visited the menagerie, and had pricked the extremity of the trunk of the creature with some sharp instrument, taking care in doing so to be beyond its reach.

101. Even fishes do not appear to be altogether destitute of memory, since eels approach upon the call of their keeper. Serpents in menageries also manifest the same faculty.

102. The actions by which animals show the exercise of a certain degree of reasoning are scarcely less numerous. Thus, the dog, which is kept in a cage, will gnaw the bars if they are of wood, but will quietly resign himself to his captivity if they are of iron, because he understands that since he can make an impression on the bars in the first case by gnawing them, he may by continued efforts cut them through and effect his liberation; but finding the first efforts in the other case unavailing, he infers that their continuance could never accomplish his object.

When a dog sees his master put on his hat, the animal infers at once that he is going out, and jumping upon him loads him with caresses to induce his master to take him as his companion. In this case there is reasoning, comparison, judgment, and a certain degree of generalisation. The dog *generalises* the act of putting on the hat, and *infers* its consequences, he *remembers* the act done on former occasions, and that it was followed by a walk abroad on the part of the master, and he *concludes* that what took place before will under like circumstances occur again.

103. A watch-dog, which was habitually chained to his box, found that his collar was large enough to allow him to withdraw his head from it at will. Reflecting, however, that if he practised this manœuvre when exposed to the observation of his master or keeper, the repetition of the act would be necessarily prevented by the tightening of the collar, he refrained from practising it by day, but availing himself of the expedient by night, roamed about the adjacent fields which were stocked with sheep and lambs, some of which, on these occasions, he would wound or kill. Bearing on his mouth the marks of his misdeeds, he would go to a neighbouring stream to wash off the blood, having done which he would return to his box before daybreak, and, slipping his head into the collar, lie down in his bed as though he had been there during the night.

104. In the series of observations and experiments by which F. Cuvier demonstrated the gradually increasing share of intelligence given to mammifers, proceeding from the lowest to the highest species, he showed from observations made on the habits and manners of marmots, beavers, squirrels, hares, &c., that rodents in general do not possess even that common degree of intelligence which would enable them in all cases to recognise their master or to know each other. The limited intelligence of the ruminants was shown in the case of a bison in the menagerie of the Garden of Plants, which having learned to recognise its keeper, ceased to know him when he changed his dress, and attacked him as it would have attacked a stranger. The keeper having resumed his original costume, was instantly recognised by the animal.

Two Barbary rams, which occupied the same stall, having been shorn, ceased to recognise each other, and immediately engaged in battle.

105. The manners of the elephant and horse are in obvious accordance with the rank assigned to them by Cuvier in the order of intelligence. But the pig species might seem at the first more doubtful. Nevertheless, Cuvier found that it was very little inferior to the elephant in sagacity. He found that the pecari, or South American hog, was as docile and familiar as the best trained dog. The wild boar is easily tamed,

158

recognises and obeys his keeper, and is capable of learning certain exercises.

106. The increasing degree of intelligence ascending from the Carnivora to the Quadrumana was clearly established by the observations of Cuvier, who found that in accordance with his system, the ourang-outang, of all mammifers, manifested the highest degree of intelligence.

107. A young ourang-outang, of the age of fifteen or sixteen months, was an especial object of observation and experiment. He showed the greatest desire for society, manifesting the strongest attachment for those who had charge of him. He loved to be caressed by them, and used not only to embrace, but even to kiss them. He pouted like a child when not allowed to have his way, and testified his vexation by cries, rolling himself on the ground, and striking his head upon it, so as to excite compassion by hurting himself.

This animal used to amuse itself by climbing up the trees in the Garden of Plants, and perching on their branches. It happened one day, that the keeper attempted to climb the tree to catch it. The ourang-outang immediately shook the tree with all its force, so as to deter the keeper from mounting it. The keeper then retired, and after an interval returned, approaching the tree, when the ourang-outang again set itself to shake the branches. "In whatever manner," says Cuvier, "this conduct may be viewed, it will be impossible not to see in it a combination of ideas, and to recognise in the animal capable of it the faculty of generalisation."

In fact, the ourang-outang in this case evidently reasoned by analogy from himself to others. He had already experienced the alarm excited in his own mind by the violent agitation of the bodies on which he was supported. He argued, therefore, from the fear which he felt himself to the fear which others would suffer in like circumstances. In other words, as Cuvier justly observes, he erected a general rule upon the basis of a particular circumstance.

This animal being one day shut up alone in a room, it availed itself of a chair which happened to be placed at the door, upon which it mounted to reach the latch. To prevent this manœuvre the keeper removed the chair; but the animal, when he had departed, seized another chair which was at a distance from the door, and placing it under the latch, mounted upon it in like manner.

In this case we find all the indications of memory, judgment, generalisation, and reasoning. The case is totally different from those so frequently witnessed in the case of animals trained for exhibition. The animal had never been taught to mount upon a chair to reach the latch of the door, nor had he ever seen any one do so. It must therefore have been by his own experience alone that he learned to perform the act. By observing the actions of his keepers, he learned that chairs could be removed from one place to another. Generalising this, he inferred that he could remove a chair to the door. He learned also by his own experience, that by mounting on chairs and tables, he could reach objects which were unattainable from the floor, and, generalising this experience, inferred that he could by the same expedient reach the latch.*

It is impossible in cases like these to admit instinct as an explanation of the phenomenon. The circumstances under which such acts are performed, and the consequences which attend them, are incompatible with all the conditions usually attached to the faculty of instinct.

* Milne Edwards's Zoology, p. 256.

FIG. 27.—OURANG-OUTANG.

INSTINCT AND INTELLIGENCE.

CHAPTER IV.

108. THE ourang-outang has been a subject of observation with all naturalists who have devoted their labours to the investigation of the habits of animals.

Buffon records circumstances respecting this animal that places him in close relation with man. Thus he has seen him present his hand to visitors to conduct them to the door, walk gravely with them as a friend or companion would, sit at table and spread his napkin in a proper manner, and wipe his lips with it, use a

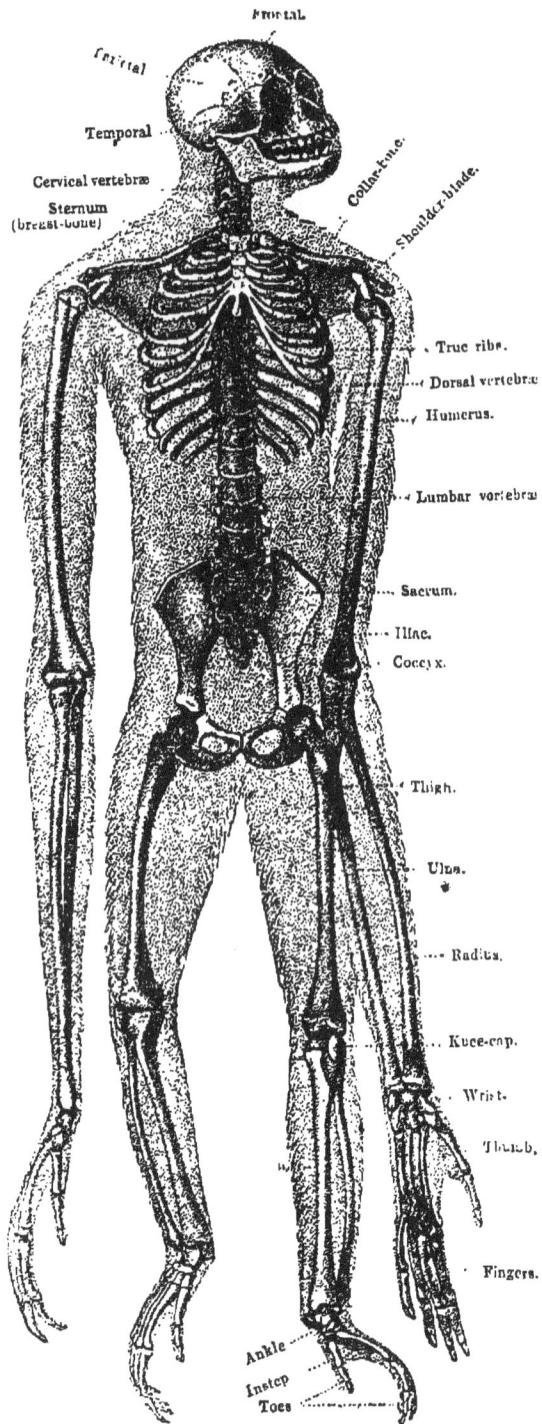

Frontal

Parietal

Temporal

Cervical vertebræ

Sternum
(breast-bone)

Collar-bone.

Shoulder-blade.

True ribs.

Dorsal vertebræ

Humerus.

Lumbar vertebræ

Sacrum.

Iliac.

Coccyx.

Thigh.

Ulna.

Radius.

Knee-cap.

Wrist.

Thumb.

Fingers.

Ankle

Instep

Toes

1'2   Fig. 28.—Skeleton of Ourang-Outang.

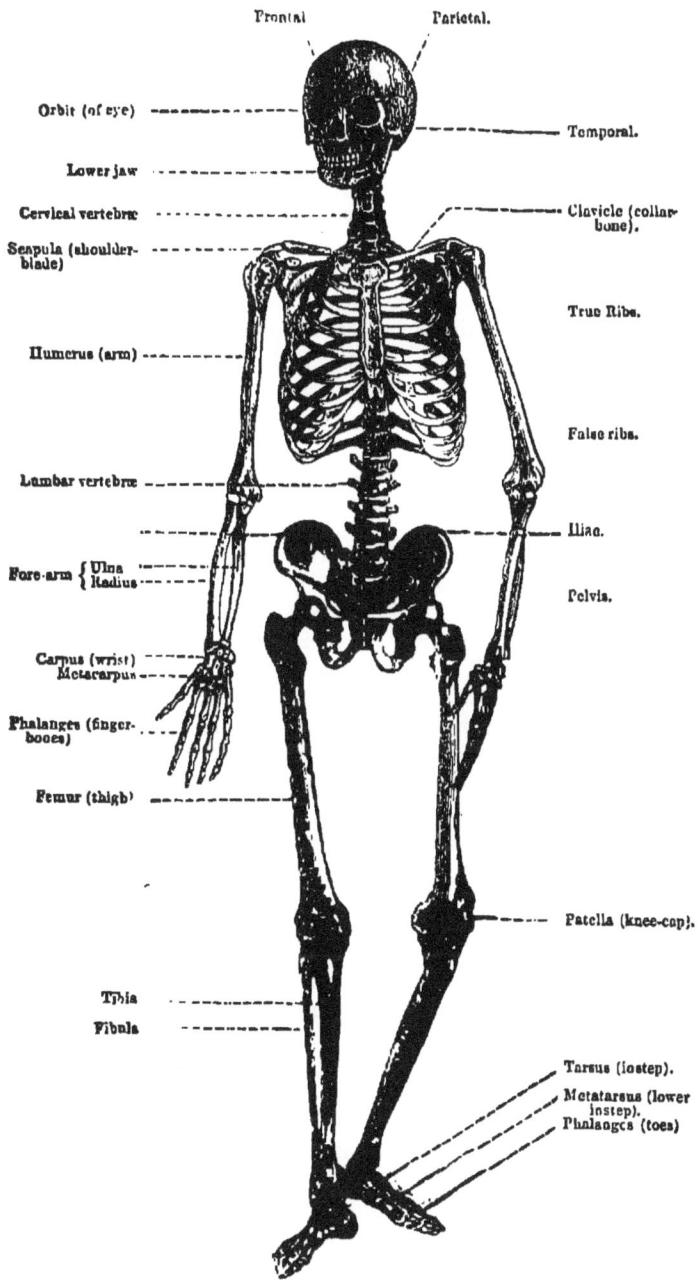

Frontal.   Parietal.

Orbit (of eye)

Temporal.

Lower jaw

Cervical vertebræ

Clavicle (collar-bone).

Scapula (shoulder-blade)

True Ribs.

Humerus (arm)

False ribs.

Lumbar vertebræ

Iliac.

Fore-arm { Ulna
       Radius

Pelvis.

Carpus (wrist)
Metacarpus

Phalanges (finger-bones)

Femur (thigh)

Patella (knee-cap).

Tibia

Fibula

Tarsus (instep).

Metatarsus (lower instep).
Phalanges (toes)

Fig. 20.—Human Skeleton.

M 2

spoon and fork to convey food to his mouth, pour wine into a glass and drink it, take wine with another at the table when so invited, clinking the glass according to the French custom; he would go and fetch a cup and saucer, put them on the table, put sugar in the cup, pour tea into it, and leave it to cool before drinking it, and all this without any prompting on the part of the master. He was circumspect in approaching persons, to avoid the appearance of rudeness, and used to present himself like a child desirous of receiving caresses.

M. Flourens found the same marks of intelligence in an ourang-outang in the Garden of Plants. This animal was gentle and sensible to caresses, especially from children, with whom he was always delighted to play.

He could lock and unlock the door of his room, and would look for the key of it. He showed none of the petulance and impatience common to apes. His air was serious, his gait grave, and his movements measured.

It appeared one day that an illustrious old savant accompanied M. Flourens to visit the animal. The figure and costume of this gentleman were singular. His body stooped, his gait was feeble, and movement slow. These peculiarities evidently attracted the notice of the animal. While he acquiesced with all that was desired of him, his eye was never withdrawn from his strange visitor. When they were about to retire, the animal, approaching the old gentleman, took with a certain expression of archness the cane from his hand, and affecting to support himself upon it, bent his back and hobbled round the room, imitating the gait and gestures of the stranger, after which, with the greatest gentleness, he returned to him the walking-cane.

"We quitted the ourang-outang," says M. Flourens, "convinced that philosophers are not the only observers in the world."

109. The close analogy of the structure of the ourang-outang to that of man will render this high degree of intelligence less surprising. This analogy is even more apparent in the skeleton than in the mere external form, as will be seen by comparing the fig. 28, which is that of the ourang-outang, with fig. 29, which is that of man.

110. An analogy not less striking is apparent in the brain of the animal compared with the human brain. In fig. 30 a side view of the human brain is presented, and in fig. 31 a similar view of the brain of the ourang-outang.

111. Leroy had already observed in the wolf, like signs of generalisation. When that animal appears, he is pursued, and the assemblage and tumult announce to him at once how much he is feared, and all that he has himself to dread. Hence, when-

164

ever the scent of man strikes his sense, it awakens in him the

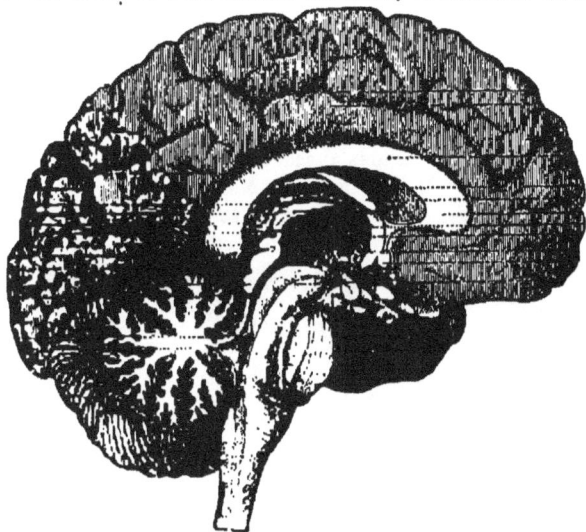

Fig. 30.—Human Brain.

idea of danger. While this fearful accessory attends it, the

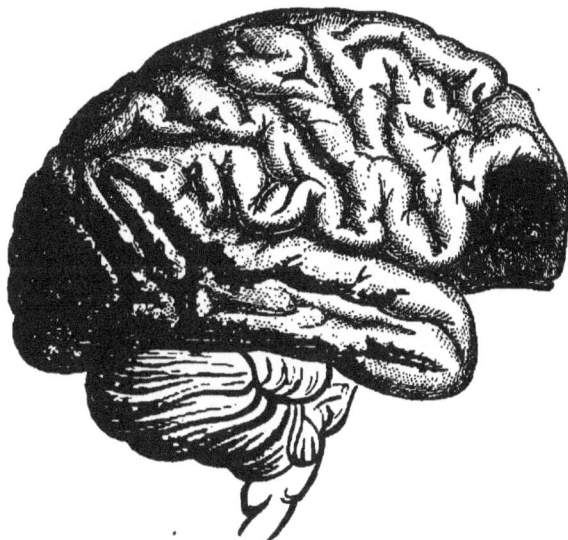

Fig. 31.*—Brain of the Ourang-Outang.

* This figure is slightly incorrect. The brain of the ourang does not quite overlay the cerebellum.

165

most seductive prey will not attract him; and even when the cause of danger is not present, the desired object is long regarded with suspicion. The wolf therefore, observes Leroy, must necessarily have an abstract idea of the danger, since he cannot be supposed to have a knowledge of the snares which are spread for him on any particular occasion.

112. The following curious anecdote of the habits of hawks and falcons is related by M. Dureau de la Malle.*

These birds, when they return from the pursuit of their prey

at the season when their younglings have become sufficiently fledged to rise on the wing, bring back in their talons some object, such as a mouse or sparrow, which they have killed, for the purpose of giving a lesson to their young in the art of capturing their prey. These birds are observed to have peculiar calls, which their young understand, and which are always repeated for the same purpose. M. de la Malle, who had a lodging in the Louvre, observed one day a male and female falcon thus returning and bringing with them a dead sparrow in their talons. They soared in the air over their nest, calling their younglings with the cry intended to summon them to rise on the wing. When the young birds thus rose, the old ones,

* Mémoire sur le développement des facultés intellectuelles des Animaux.

166

soaring vertically over them, let fall the sparrow, upon which the younglings pounced. In the first attempts, the latter invariably failed in seizing the sparrow, not being yet sufficiently adroit. The old birds would then descend, and, seizing the prey, rise with it into the air once more, and let it fall again upon the young; nor would they allow the latter to devour it until they succeeded in catching it as it fell.

These lessons were progressive. The prey first let fall on their younglings was dead. When they had acquired sufficient skill to seize this in falling through the air, the old parents brought living birds, first more or less disabled, and afterwards uninjured, upon which they exercised their young in the same manner; and this was continued until the young birds were fully able to pursue and seize their prey without further practice or instruction.

Every one has seen the cat give to her kittens similar progressive lessons.

She commences by biting a mouse so as to stun, or slightly disable, without killing it. She then liberates this mouse before her kittens, and encourages them to pursue it, the matron cat standing by, a vigilant observer of the scene. If the mouse shows any sign of escaping, she immediately pounces upon it, and disables it so effectually, that her kittens soon finish it.

According to Daubenton, the eagle carries its eaglet aloft upon its wings, and letting it go in mid air, tries its powers of flight. If its strength fails, the mother is sure to be at hand to support it.

113. Among the acts of animals which are obvious results of intelligence and not of instinct, the following may serve as instructive and interesting examples:—

Plutarch relates, that a dog desiring to drink the oil contained in a pitcher with a narrow mouth, the surface of the liquid being so low as to be out of the reach of his mouth, threw pebbles into it, which sinking in the oil, caused its surface to rise so high that the dog could lap it up. According to Plutarch, the dog must in this case have reasoned thus: the pebbles being heavier than the oil will sink to the bottom; they will displace part of the oil, and will displace more and more the more of them that are thrown in; therefore by throwing in a sufficient number, the surface of the oil must necessarily rise to the dog's mouth.

114. M. Flourens relates the following anecdote of bears in the Garden of Plants:—

It happened that these animals multiplied until there were more of them than it was desired to keep, and it was resolved to get rid of two. It was proposed to poison them with prussic acid. For this purpose some drops of that liquid were poured upon little cakes, which being offered to the bears in the usual way, the animals stood up on their hind legs, and opened their mouths to catch them. The moment they received them, however, they spat them out, and retired to a remote corner of their den, as though they were frightened. After a short interval, however, they returned to the cakes, and pushed them with their paws into the water-trough left to supply them with drink, and there they carefully washed them by agitating them to and fro in the water. After this they smelled them, and again washed them, and continued this process until the poison was washed off, when they ate the cakes with impunity. All the poisoned cakes given to them were thus treated, while all the cakes not poisoned were devoured immediately.

The animals which had shown these singular marks of intelligence were spared the fate to which they had previously been condemned.

115. One of the most remarkable circumstances attending the faculty of intelligence, observed not only in the ourang-outang, but in all species of apes, is that its greatest development is manifested when the animal is young, and that instead of improving, it decreases rapidly with age. The ourang-outang when young excites surprise by his sagacity, cunning, and address. Having attained the adult state, he is a gross, brutal, and intractable animal.* In this, as well as in all other species of apes, the decrease of intelligence is commensurate with the increase of growth and strength. The intelligence of the animal, therefore, such as it is, is not like that of man, perfectable.

116. It is established, therefore, by the observations and researches of naturalists, that intelligence is a faculty common to man and to inferior animals. According to some, man is distinguished from other animals only by the degree in which he is endowed with this faculty; and the difference of degree is so immense, that, before accurate observations had proved the contrary, the faculty of intelligence was deemed the exclusive gift of the human race. Others contend that the intelligence of man differs from that of animals not in *degree* only, but in *kind ;* that, in short, what is called intelligence in animals, is a faculty essentially different from what is called intelligence in man, and ought to have been called by a different name.

The intelligence of animals is limited and stationary. It is unimproveable and incommunicable. The intelligence of man, on the contrary, is susceptible of improvement without limit, and

* Flourens, "De l'Instinct et de l'Intelligence des Animaux," p. 35.

may be imparted from individual to individual. It radiates like light. Its power of growth and improvement is indefinite.

As we observed before, much of the obscurity and confusion which has attended all discussion respecting the intelligence of animals, arose from the omission of a sufficiently clear line of demarcation between instinct, properly so called, and intelligence.

The great purposes of instinct are the preservation of the individual and the continuance of the species. To plants, which live and die without change of place, the Creator has given strong and elastic tissues to ensure the preservation of the individual, and myriads of germs are put in immediate juxta-position with the organs destined to fecundate them, to ensure the continuance of the species.

To animals, which are endowed with powers of locomotion, and which are thereby exposed to numerous vicissitudes, God has given instinct to preserve the individual, to reproduce the species, and to perpetuate His work, thus rendering them unconscious agents in fulfilling His almighty command to "increase and multiply."

Instinct is then a gift emanating direct from divine goodness, and being a gift, and not a faculty, is inexplicable. It is a power inseparable from animal life. Its dictates are as imperious as those of gravitation or magnetism. It can neither be modified nor evaded. The bee constructs her comb in one manner and on one plan, from which no bee, old or young, ever. departs. The bird builds its nest after a fashion as uniform, and by a law as rigorous, as that by which the lilies of the field put forth their blossoms.

Nor is man himself more emancipated from the sway of instinct. His first act on coming into the world is the instinctive seizure of the maternal nipple. Fear is the instinct of self-preservation; love that of the continuance of the species.

Intelligence on the one hand is the power of comprehending the consequences of acts, and of giving to them a direction determined by the will of the agent.

Reason is the most exalted form of intelligence, so exalted that some contend that it ought to be considered as a distinct faculty. It is by reason that man knows himself, judges himself, and conducts himself.

Animals are variously gifted with intelligence, for they are endowed with perception, memory, and consciousness. They are susceptible of passions and affections, not only physical, but moral. All the human passions, anger, hatred, jealousy and revenge, agitate them. They are devoted, affectionate, grateful, prudent, circumspect, and cunning. Kindness soothes and melts them. Injury awakens their resentment. The movements of the brain, like those of the human encephalon, evokes in sleep their waking thoughts and desires. The dog of the chace dreams that he pursues the hare, and the more peaceful follower of the shepherd, that he collects the straying flock.

The intelligence of animals is rigorously limited to the objects of the external world that are presented to their senses. The intelligence of man has a far wider range. By the senses it is put in relation with the material world; by consciousness, with the inner being, the soul, and by intuitive ideas and sentiments with God.

The exalted intelligence of man confers on each individual a character as distinct as his features. He acquires from it his peculiar habits, qualities, tendencies, virtues, and faults. While it makes him free in one sense, it isolates him in another. Instinct, on the contrary, effaces individual distinction,—reducing all to a common type. All beavers, and all bees, lead lives absolutely alike, and may be regarded as differing no more than the units which make up an abstract number.

117. The inferior animals are endowed, as we have seen, largely with the powers of sensation, perception, and memory. They also possess, though in a very inferior degree, powers of comparison, generalisation, judgment, and foresight. In what then, it may be asked, consists the mark of the vast difference in degree of their intelligence, as compared with the mental powers exercised by the human race. This question has been satisfactorily answered by the observations and researches of Frederick Cuvier, Flourens, and others. According to these physiologists, animals receive by their senses impressions similar to those which are received by ours. Like us, also, they preserve and are able to recall the traces of these impressions. And such perceptions being thus preserved, supply for them as for us numerous and varied associations. Like us they combine them, observe their relations, and deduce conclusions from them, and to this extent, but not beyond it, their intelligence goes; but they have not a glimpse of that class of ideas which Locke denominates ideas of reflection. These, as is well known, are the perceptions which man acquires, not by his organs of sense, but by the power with which he is endowed to render his mind itself, and its operations, the subjects of contemplation and perception. Man has as clear a perception of the faculty of memory, for example, as he has of the colours of the rainbow. The scent of a rose is not more distinct to his apprehension than are his mental powers of comparison and induction. In short, his ideas of reflection are as vivid and definite as his ideas of sensation, and may, indeed, be said to be even more permanent and inseparable from his intellectual existence. He may be deprived of one or more of his

170

organs of sense, and thus cease to have any perception of the qualities peculiar to that organ, save those which his memory may supply. But so long as he exists and thinks, nothing can deprive him of the immediate perception of the ideas of reflection.

Of this class of ideas there is not the slightest trace in the inferior animals, and herein lies the line of demarcation which separates the human race from them, and places it immeasurably above them. Animal intelligence never contemplates itself, never sees itself, never knows itself. It is utterly incapable of that high faculty by which the mind of man, as Locke observes, "turns its view inward upon itself." That thought which contemplates itself; that intelligence which sees itself, and studies itself; that knowledge which knows itself, constitutes a distinct order of mental phenomena to which no inferior animal can attain. These constitute, so to speak, the purely intellectual world; and to man alone, here below, that world belongs. In a word, the animals feel, know, and think; but to man alone of all created beings it is given to feel that he feels, to know that he knows, and to think that he thinks.

118. Of all the instruments by which the range of intelligence is enlarged, and the power of reason augmented, language is assuredly the most important. It is the means by which feelings are expressed and knowledge imparted. It is the instrument by which the observation and experience of individuals is rendered common property.

Language, in the only sense in which it is an instrument of intelligence, is not the mere mechanical production of distinct sounds by the vocal organs, for in this sense parrots may be said to be endowed with it. It is a divine gift and not a faculty. Its origin has been sought for by the learned, but sought in vain. Like the instinct of self-preservation and reproduction, it has been an immediate emanation of divine power. God made it as he made light. He said, "Let man speak," and man spoke!

Most animals have voice, but man alone has language. It is by language, more than any other external character, that man is distinguished. The animals which come nearest to him in their physical organisation, such as the ourang-outang and other apes, are as completely deprived of language as those which are most removed from him. Man is thus separated from the lower animals by a bottomless abyss.

So important is language, as a means of extending the intelligence, that in a moral sense it may be said, that to speak or not to speak, is to be or not to be!

There can be no doubt in the mind of any careful observer, that the chief obstacle to the extension of the natural intelligence of many animals is the want of language to express their feelings and thoughts. It is evident that if the dog or the ourang-outang, which was the subject of Cuvier's experiments, could speak, their intelligence would be vastly enlarged.

Deprived of language, the more intelligent of the inferior animals seem, like the dumb, deeply conscious of the want, and make supernatural efforts to supply it and to make their sentiments understood. For this purpose they resort to ingeniously modulated vocal sounds, to signs and gestures. Each creature invents for itself a sort of pantomimic and highly expressive language. The dog appeals to you by gently laying his paw upon you, and if that fail to awaken your attention, he strokes you or taps you with it, as if he knew that you would thus be more apt to *feel* his solicitation. Does the cat desire to have some want supplied? she raises her back and passes her soft fur in contact with your legs, and repeats the application by going round and round you. The horse waiting at your door, fresh from his stall, and impatient for air and exercise, expresses his desire by pawing the ground with his fore-foot. In the pairing season, the male bird tries to fascinate his gentle mate by spreading out the fine hues of his plumage, making circuits, and fluttering around her.

All animals that have voice at all, use its modulations as a means of expression, and render it manifest that they would speak if they could. Many and ingenious are the artifices which they use as a substitute for the admirable instrument of intercommunication with which man has been gifted.

119. Thus, for example, in the case of such mammifers and birds as usually assemble in herds or flocks, individuals are observed who, being placed as sentinels, warn their companions of the approach of danger.

Marmots and flamingoes present examples of this. It is also observed with swallows, who, when their young are menaced by an enemy, immediately call together, by their cries of distress, all the swallows of the neighbourhood, who fly to the aid of their fellows, and unite to harass the animal whose attack they fear.

120. It has been well ascertained that various species of insects have means of intercommunication. The observations of Huber, Latreille, and other naturalists, leave scarcely a doubt on this point. Thus, for example, when an ant's nest has suffered any local disturbance, the whole colony is informed of the disaster with astonishing rapidity; no appreciable sound is heard, but the particular ants who are witnesses of the fact, are seen running in various directions among their companions. They bring their heads into contact, and unite their antennæ as two persons

would who take each other by the hand. All the ants who are thus addressed are immediately observed to change their route if they were moving, and to abandon their occupation if they were at work, and to return with those from whom they received the information, to the scene of the disaster, which is soon surrounded by thousands of these insects, thus brought from a distance.

121. In the wars which the population of two neighbouring ant-hills wage with each other, scouts and outposts precede the main body of the enemy, who often return to the leaders, giving them information, the consequence of which is, a total change in the order of march. In cases where these conflicts become doubtful, and that an army finds itself in danger of defeat, the leaders are often seen to detach aides-de-camp, or orderly officers, who return in all haste to their ant-hill, to bring up reinforcements, which assemble without delay, and march to join the main body of the army.

122. Large as the range of action is, which admits of explanation either by intelligence or instinct, or by the combination of both these faculties, some acts still remain of an extraordinary character which cannot be thus explained, and which would seem to imply the existence of some faculty in certain species of inferior animals of which man is totally destitute.

123. Among these may be mentioned the curious power with which certain birds, such as pigeons and swallows, are endowed; who, after being transported in close boxes to many hundred miles from their nest, take flight upon being liberated, and without the least hesitation direct their course towards the place from which they had been taken, with a precision as unerring as if it were actually within their view. In the case of dogs, and other mammifers, who having been brought to a great distance from home find their way back, the act is explained by the extreme delicacy of their sense of smell; but no such explanation will be admissible in the case of carrier-pigeons, who, having been brought, for example, from London to Berlin, and being liberated at the latter city, instantly direct their course back to the former, flying over the great circle of the earth which joins the two places. We are not aware that any attempt has been made to refer this class of facts to any recognised faculty.

124. Closely connected with instinct and intelligence is the capability of animals to be tamed and domesticated.

Naturalists agree generally that the animals which are domesticated with greatest facility are those which in the wild state live in troops or societies. To this there is scarcely a well-established exception. The cat and the pig are apparent exceptions, but it is contended that they are never domesticated in the true sense of the term. Every one is familiar with the difference between the domestic state of the cat and that of the dog. The latter is domesticated in the truest sense of the term.

A careful distinction must be maintained between the state of *tameness* and that of *domesticity*. The species of animals which are susceptible of these states are wholly different.

Domesticity descends from the parent to the offspring. Slavery here descends as an heritage.

Tameness is produced in the individual by the immediate treatment of man. The offspring of a tame bear would be as wild as the parent was before its subjugation.

The young of tame animals must, like the parents, be tamed. The young of domestic animals are already domestic.

Gregarious animals, endowed as they are with the instinct of sociability, select by common consent a chief, to whom they yield obedience. In the domesticated state, man taking the place and exercising the influence of that chief, receives the same instinctive obedience. Domesticity is, therefore, an animal instinct, of which man avails himself to attract into his service animals of the sociable species.

Tameness, on the contrary, is not an instinct but a habit. It is produced originally by fear, and maintained by the creation of artificial wants which man alone can satisfy.

Frederick Cuvier relates an incident which strikingly illustrates the distinction between the true instinct of sociability and the fictitious state of tameness produceable by habit.

A lioness, in the menagerie of the Garden of Plants, had been reared in the same cage with a dog. The two animals became familiar friends, and a mutual attachment was manifested. The dog having died, was replaced by another, which the lioness readily enough accepted and adopted, appearing to suffer nothing from the loss of her old friend and companion. In the same manner she survived the second dog, showing no signs of grief, and as readily as before received a third dog, with whom she continued to associate in the same manner. This third dog, however, outlived the lioness. When the latter died, a touching spectacle was presented. The poor dog refused to leave the cage in which the body of his friend lay. His melancholy increased from day to day. The third day he refused all food, and on the seventh died.

The agencies by which man first tames and later domesticates animals are few and obvious. They consist chiefly of the alternate privation and satisfaction of their physical appetites, and

174

especially of those fictitious appetites which man himself excites and creates, and which he alone can gratify.

Hunger holds the first and most important place among these. It is by playing upon this appetite that the horse is tamed and trained. But little food is given at a time, and even that only at long intervals. The animal, ignorant that he who tends him is the cause of his privation, has full knowledge that it is by him that this privation is relieved, and if some choice aliment exciting to the palate is occasionally supplied, the authority of the master is augmented, and the gratitude and affection of the animal strongly awakened. It is by certain dainties, and especially by sugar, judiciously supplied and withheld, that the horses of the circus are brought to perform feats which create such general astonishment.

Privation of sleep is an agent of subjugation even more potent than hunger; and it is by hunger, pushed to excess, by the application of the whip, by stunning and alarming noises, such as those of the drum, and certain wind instruments, that this forced wakefulness is maintained.

By such means the urgent wants of the animal are excited; the power of the master is, however, acquired, not by the wants themselves, but by exhibiting himself in the most unmistakeable manner to the suffering creature as the agent of its relief. Not satisfied with presenting himself as the agent for the relief of real physical wants, he artfully creates fictitious ones, not only physical but moral. Choice food is now and then given, which none but the master can supply; but besides this the animal is rendered sensible to caresses, and after a time becomes most grateful for them. The elephant, the horse, and the cat are passionately sensible of the kindness of those with whom they are domesticated, but it is over the dog, more than any other, that the sway of this moral power extends.

A female wolf, in the Garden of Plants at Paris, became so sensitive to the caresses of its keeper, that it testified a delirium of joy at the sound of his voice or the touch of his hand. A Senegal jackal betrayed like emotions excited by a similar cause, and a common fox was habitually so affected by the caresses of its keeper that it was found necessary to discontinue such excitement.*

The process of subjugation of the wild animal is then one which attains its object by address and seduction. Natural wants are made to be felt, and fictitious ones are created, that man may have the merit of supplying them. He thus renders himself more

* Mem., Fred. Cuvier.

and more necessary by the benefits he confers; and having arrived at that point, he ventures to employ fear and chastisement, which if resorted to without the previous measures would have excited resistance and repugnance.

To tame an animal is not to train him. Tameness is the subjugation of those instincts which would render him hurtful to those around him. Training is directed to the intelligence rather than the instinct. It is an educational process, which develops intelligence while it weakens instinct. Savages, while they are less intelligent than the civilised, have surer and quicker instincts. It is the same with the lower animals. Domesticity always enfeebles and often wholly effaces instinct.

When man educates and trains an animal, he imparts to it a ray of his own intelligence. The change is rather that of a new faculty created than of an existing one enlarged. It is a transformation rather than an improvement.

www.ingramcontent.com/pod-product-compliance
Lightning Source LLC
Chambersburg PA
CBHW021704210326
41599CB00013B/1515